International Conference on Research and Applications of Plasmas (PLASMA-2017)

Journal of Physics: Conference Series Volume 959

Warsaw, Poland
18 September 2017

ISBN: 978-1-5108-5850-3
ISSN: 1742-6588

Printed from e-media with permission by:

Curran Associates, Inc.
57 Morehouse Lane
Red Hook, NY 12571

Some format issues inherent in the e-media version may also appear in this print version.

Printed by Curran Associates, Inc. (2018)

For permission requests, please contact the Institute of Physics
at the address below.

Institute of Physics
Dirac House, Temple Back
Bristol BS1 6BE UK

Phone: 44 1 17 929 7481
Fax: 44 1 17 920 0979

techtracking@iop.org

Additional copies of this publication are available from:

Curran Associates, Inc.
57 Morehouse Lane
Red Hook, NY 12571 USA
Phone: 845-758-0400
Fax: 845-758-2633
Email: curran@proceedings.com
Web: www.proceedings.com

Table of contents

Volume 959

International Conferences on Research and Applications of Plasmas, PLASMA-2017

18 September 2017, Warsaw, Poland

Accepted papers received: 17 January 2018
Published online: 7 February 2018

Preface

International Conferences on Research and Applications of Plasmas, PLASMA-2017

Peer review statement

Papers

Invited Presentations

Laser-driven ion acceleration: methods, challenges and prospects

J Badziak.....1

Oral Presentations

Studies of runaway electrons via Cherenkov effect in tokamaks

J Zebrowski, L Jakubowski, M Rabinski, M J Sadowski, M J Jakubowski, R Kwiatkowski, K Malinowski, R Mirowski, J Mlynar, O Ficker, V Weinzettl, F Causa, COMPASS and FTU Teams.....16

Influence of gas conditions on electron temperature inside a pinch column of plasma-focus discharge

D R Zaloga, M J Sadowski, E Skladnik-Sadowska, M Paduch, E Zielinska and K Tomaszewski.....24

Materials surface damage and modification under high power plasma exposures

I Garkusha, V Makhlaj, O Byrka, V Taran, V Voitsenya, S Malykhin, S Herashchenko, S Surovitskiy, K Nowakowska-Langier, M J Sadowski, E Skladnik-Sadowska and D. Terentyev.....30

Strong electromagnetic pulses generated in high-intensity laser-matter interactions

P Rączka, J-L Dubois, S Hulin, M Rosiński, A Zaraś-Szydłowska and J Badziak.....38

Posters

Analysis of optical spectra from steel samples exposed to pulsed plasma streams

A Marchenko, I Garkusha, V Makhlai, E Skladnik-Sadowska, R Kwiatkowski, K Nowakowska-Langier, M J Sadowski and M Kubkowska.....45

Study of Plasma Behavior during ECRH Injection in the GAMMA 10 SMBI Experiments

Md. Maidul Islam, Yousuke Nakashima, Shinji Kobayashi, Nobuhiro Nishino, Kazuya Ichimura, Takaaki Iijima, Md. Shahinul Islam, Takayuki Yokodo, Guanyi Lee, Tsubasa Yoshimoto, Sotaro Yamashita, Masayuki Yoshikawa, Junko Kohagura, Mafumi Hirata, Ryutaro Minami, Tsuyoshi Kariya, Ryuya Ikezoe, Makoto Ichimura, Mizuki Sakamoto and Tsuyoshi Imai.....49

The preparation and analysis of ceramic fibers with deposited catalytic particles by plasma technology

M Milieška, R Kėželis and V Snapkauskienė.....53

Molecular ND Band Spectroscopy in the Divertor Region of Nitrogen Seeded JET Discharges

E Pawelec, T Dittmar, A Drenik, A Meigs and JET contributors.....57

Proton acceleration in two-species targets irradiated by an ultra-intense femtosecond laser pulse

J Domański, J Badziak and S Jabłoński.....61

Fixation of the stressed state of glass plates by coating them with thin films using a plasma focus installation

V.N. Kolokoltsev, V.F. Degtiarev, I.V. Borovitskaya, V.Ya. Nikulin, E.N. Peregudova, P.V. Silin and A.A. Eriskin.....65

Generation of sub-gigabar-pressure shocks by a hyper-velocity impact in the collider driven by laser-induced cavity pressure

J Badziak, M Kucharik and R Liska.....70

IOP Conf. Series: Journal of Physics: Conf. Series **959** (2018) 011001 doi:10.1088/1742-6596/959/1/011001

The International Conferences on Research and Applications of Plasmas (PLASMA) have been organized every two years since 1993. The scope of the PLASMA conference series, which idea was originally to facilitate the exchange of knowledge and to spread the collaboration between the scientists and engineers from the Eastern European Countries, covers most of issues of broadly understood plasma physics and technology.

PLASMA-2017, organized by the Institute of Plasma Physics and Laser Microfusion (IPPLM), took place from 18th September to the 22nd September 2017 at the Central Agriculture Library in Warsaw, Poland. The conference covered scientific areas like:

I. Elementary processes and general plasma physics

II. Plasmas in tokamaks and stellarators (MCF)

III. Plasmas generated by laser beams and Inertial Confinement Fusion (ICF)

IV. Plasma produced by Z-pinch and Plasma-Focus discharges

V. Low-temperature and dusty plasma

VI. Space plasmas and laboratory astrophysics

Peer review statement

All papers published in this volume of *Journal of Physics: Conference Series* have been peer reviewed through processes administered by the proceedings Editors. Reviews were conducted by expert referees to the professional and scientific standards expected of a proceedings journal published by IOP Publishing.

IOP Conf. Series: Journal of Physics: Conf. Series **959** (2018) 012001 doi:10.1088/1742-6596/959/1/012001

Laser-driven ion acceleration: methods, challenges and prospects

J Badziak

Institute of Plasma Physics and Laser Microfusion, 01-497 Warsaw, Poland

E-mail: jan.badziak@ifpilm.pl

Abstract. The recent development of laser technology has resulted in the construction of short-pulse lasers capable of generating fs light pulses with PW powers and intensities exceeding 10^{21} W/cm^2, and has laid the basis for the multi-PW lasers, just being built in Europe, that will produce fs pulses of ultra-relativistic intensities ~ 10^{23} – 10^{24} W/cm^2. The interaction of such an intense laser pulse with a dense target can result in the generation of collimated beams of ions of multi-MeV to GeV energies of sub-ps time durations and of extremely high beam intensities and ion fluencies, barely attainable with conventional RF-driven accelerators. Ion beams with such unique features have the potential for application in various fields of scientific research as well as in medical and technological developments. This paper provides a brief review of state-of-the art in laser-driven ion acceleration, with a focus on basic ion acceleration mechanisms and the production of ultra-intense ion beams. The challenges facing laser-driven ion acceleration studies, in particular those connected with potential applications of laser-accelerated ion beams, are also discussed.

1. Introduction

High-energy particle accelerators driven by a radio-frequency (RF) electromagnetic field are among the largest and most complex facilities built on Earth. Currently, the largest RF-driven accelerator is the Large Hadron Collider (LHC) at CERN, which accelerates protons to multi-TeV energies (1 TeV = 10^{12} eV) in a 27-km high-vacuum ring. One of the main reasons for the enormous size of the accelerator is a limit on the acceleration field strength, which is about 1 MV/cm. Due to this limit, to accelerate ions/protons to multi-GeV or TeV energies, the acceleration path has to be very long, meaning that the accelerator has to be very large. This lead to the question as to whether RF-driven ion accelerators are the only possible ones that enable us to accelerate ions to very high velocities close to the speed of light c. One potential alternative, or at least an important supplement to the RF-driven ion accelerators, seems to be ion accelerators driven by an intense laser beam.

A simple scheme of a laser-driven ion accelerator is shown in Figure 1. The accelerator consists of a short-pulse high-intensity laser and a target, e.g. thin foil, placed in a vacuum chamber. The laser beam interacting with the target produces plasma, in which electrons are partly separated from ions by the action of the laser field. Between the layer of electrons and the ions, a very strong electric field is produced. This field pulls the ions, which follow the moving electron layer. The field strength can reach extremely high values (up to tens or even hundreds GeV/cm) and, as a result, the ions can be accelerated to high energies over sub-mm distances, by many orders of magnitude shorter than required in conventional RF-driven accelerators. In this way, the laser-driven accelerator can potentially be much smaller and less complex than conventional ones. Furthermore, since the laser-accelerated ion bunch is very dense and compact, the ion beam intensities/powers can be very high, and the time duration of the ion pulse can be very short. Such ion beams have the potential to be used in various branches of science, technology and medicine, and can significantly extend the current scope of ion beam applications.

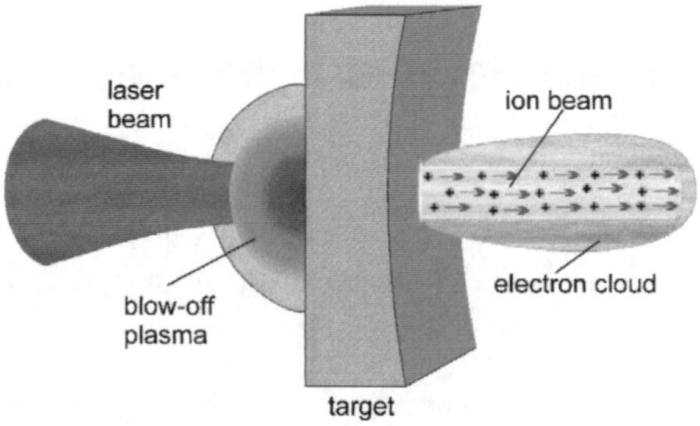

Figure 1. A simple scheme of the laser-driven ion accelerator.

Studies into the generation of energetic ions from laser-produced plasma have a long history and begin in the nineteen sixties [1,2]. However, a real breakthrough in these studies came at the turn of the millennium, when collimated forward-emitted beams of MeV protons generated from thin foil targets irradiated by a short (≤ 1 ps) high-power laser pulse were demonstrated [3-5]. Following these pioneering experiments, the research on laser-driven ion acceleration has developed rapidly and this research field is currently one of the most intensely studied areas of plasma physics.

This paper is a brief review of recent research on laser-driven ion acceleration based on a talk presented at the International Conference on Research and Application of Plasmas, held in Warsaw, in September, 2017. Interested readers can find an extended review of this topic in [6,7]. In section 2 of this paper, the forces and mechanisms of ion acceleration in laser-produced plasma are presented. In sections 3 and 4, the basic features of the most studied mechanisms of ion acceleration, namely the target normal sheath acceleration (sec. 3) and the radiation pressure acceleration (sec. 4) are described. Section 5 presents examples of numerical studies on the acceleration of protons and super-heavy ions at ultra-relativistic laser intensities performed by the IPPLM team. Section 6 discusses the main challenges faced by research on laser-driven ion acceleration, and section 7 summarises the main conclusions from the paper.

2. Basic forces and mechanisms accelerating ions in plasma

The main forces induced by a laser beam in plasma, which are capable of accelerating ions to high velocities over very short distances, can be roughly divided into the electromagnetic (EM) forces and the hydrodynamic forces. The EM forces clearly dominate in the case of high (> 10^{15} W/cm^2) and very high (> 10^{18} W/cm^2) intensities of the laser beam interacting with the plasma (the latter are often called the relativistic intensities, since at such intensities an electron placed in the laser field can reach velocities close to the speed of light c). The field strengths of the laser-induced EM forces can reach extremely high values (up to > 100 GV/cm) and, as a result, these forces can accelerate ions to high velocities v_i, including relativistic ones ($v_i \approx c$), over sub-mm distances. The laser-driven ion accelerators, in which EM forces are the main forces accelerating ions, can be referred to as *laser-driven electromagnetic accelerators.*

The basic source of the EM forces accelerating ions in plasma at high and very high laser intensities is the Lorenz force $\mathbf{F} = q\,\mathbf{E} + q\,(\mathbf{v} \times \mathbf{B})$, where q is the particle charge, **v** is the particle velocity and **E** and **B** are the electric and the magnetic part of the laser field in the plasma. This force interacts with both plasma electrons and plasma ions, though in practice (at laser intensities < 10^{24} W/cm^2) the movement of ions due to the direct action of this force can be neglected (due to a large ion mass) and only electrons are set in motion by the force. The electrons are pushed away from the ions (from the equilibrium position) by the force, which results in creating a huge electric field between the moving electron cloud and the ions. This electric field accelerates ions, which are usually moving together with the electron cloud. Since the electric field strength is extremely high, the ions can reach high velocities in very short distances and during very short time periods. A detailed mechanism of ion acceleration depends on both the laser pulse and the target parameters, and the acceleration mechanisms can be different for different laser-target interaction conditions. Presently, several ion acceleration mechanisms are known and have been extensively studied. They include: the target normal sheath acceleration (TNSA) [6,7,8-15], the skin-layer ponderomotive acceleration (SLPA) [16-20], the radiation pressure acceleration (RPA) [6,7,21-26], the laser break-out afterburner (BOA) [6,7,27,28], the collisionless electrostatic shock acceleration (CESA) [6,29,30], the ion solitary wave acceleration (ISWA) [31,32] and the Coulomb explosion acceleration (CEA) [33,34]. In a real experiment, two or even more acceleration mechanisms indicated above can contribute to the ion acceleration process.

At low laser intensities (~ 10^{10} – 10^{15} W/cm^2) and long laser pulses (from tens of ps to tens of ns), a significant or even dominant role in the process of accelerating plasma ions can be played by hydrodynamic forces (hydrodynamic or thermal pressure) induced by the laser pulse. In a real situation, these forces are usually a complex function of density n and temperature T of the plasma electrons, $F_h = f(n,T)$, and their absolute values are higher for higher values of n and T. The velocities of ions driven by the hydrodynamic forces are relatively small ($v_i < 0.01$ c), but the number and density of the accelerated ions can be very high. As a result, the total mass and kinetic energy of the accelerated ion (plasma) bunch can be high, even higher than in the case of EM ion acceleration driven by a short (sub-ps) laser pulse (attainable energies of long-pulse laser drivers are much higher than those of short-pulse laser drivers). The laser-driven ion accelerators, in which hydrodynamic forces play a dominant role in ion acceleration, can be referred to as *laser-driven hydrodynamic accelerators.* Examples of this kind of accelerator are the ablative acceleration systems [35-38] (used, among other things, for the acceleration and compression of DT fuel in fusion targets) or accelerators using the laser-induced cavity pressure acceleration (LICPA) mechanism [39-41].

It should be underlined that also in the low-intensity case dominated by the hydrodynamic acceleration, strong electric fields can exist in plasma (e.g.[42]), which are a result of the inhomogeneous distribution of density and velocity of plasma electrons. These fields can produce supra-thermal (fast) ions of multi-MeV energies [43], though the total energy and number of these ions is usually much lower than those of ions driven by hydrodynamic (thermal) pressure.

This paper is focused on laser-driven ion acceleration at high and very high laser intensities when ions are accelerated by EM forces. In the next two sections, the most studied mechanisms of laser-driven EM ion acceleration – namely TNSA and RPA – will be described.

3. Ion acceleration by the TNSA mechanism

A simplified scheme of ion acceleration by the TNSA mechanism [6,7,15] is shown in Figure 2. In this method, a short laser pulse interacting with the target front surface produces plasma and hot electrons with a temperature of $T_h \sim 0.1 – 10$ MeV. The electrons penetrate through the target and on the target rear surface they form a Debye sheath playing the role of a virtual cathode. The electric field induced by the cathode: $E_{ac} \sim T_h/e\lambda_{Dh} \sim 1 – 100$ GeV/cm ionises atoms at the rear surface and accelerates the produced ions over the distance $L_{ac} \sim 10$ - 50 μm to the energies $E_i \sim zeL_{ac}E_{ac} \sim 1 – 100$ MeV (λ_{Dh} is the Debye length and z is the ion charge state). The ions are accelerated mostly along the rear surface normal and, as a result, the angular divergence of the ion beam is relatively small.

The TNSA mechanism is effective in accelerating light ions, primarily protons [3-15]. Such ions can be produced from both insulator (e.g. plastic) and metal targets, typically ~ 2 – 20 μm thick; however, the metal target is usually more efficient, both in terms of the maximum ion energy and the ion beam quality (due to more efficient electron transport in metal). When a metal target is used, a

IOP Conf. Series: Journal of Physics: Conf. Series **959** (2018) 012001 doi:10.1088/1742-6596/959/1/012001

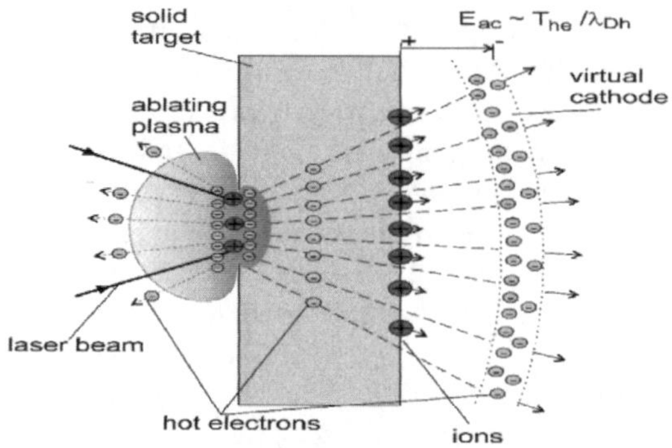

Figure 2. Idea of ion acceleration by the TNSA mechanism.

source of light ions constitute hydrocarbon contaminants (H, C) almost always existing on the target (sometimes, to increase the number of accelerated ions, the target rear surface is covered with a thin plastic layer [44]). Independent of the target composition and structure, only ions from a very thin (~ 2 – 10 nm) layer on the target rear surface are accelerated by the TNSA mechanism. As a result, the areal ion density of the ion source is relatively small, $\sigma_i < 10^{17}$ cm^{-2}, and the ion density at the source is moderate, $n_i \leq 10^{19}$ cm^{-3}. Thus, attaining high ion beam intensities $I_i = n_i v_i E_i$ (around $I_i > 10^{18}$ W/cm^2) and/or energy fluencies $F_i = \sigma_i E_i$ (around $F_i > 1$ MJ/cm^2) is possible only at very high ion energies (~ 100 MeV or higher). On the other hand, the total number of accelerated ions can be fairly high (up to ~ $10^{12} - 10^{13}$ [5-7]) since, due to the transverse transport of fast electrons in the target, the ions are extracted from an area much larger than the area of the laser beam on the target S_L (the ion source area $S_{is} >> S_L$).

The TNSA mechanism of ion/proton acceleration is fairly efficient, provided that the target rear surface is of high quality and that it is not disturbed significantly during laser-target interaction by the shock wave generated in the target by the laser pulse leading edge, or the laser pre-pulse [8,14,15]. The last requirement imposes constraints on the laser pulse duration and its contrast ratio. For relativistic laser intensities and very thin targets (the target thickness $L_T < 1$ μm), the pulse duration should be in the sub-ps range, and the contrast ratio must be high (the intensity contrast ratio > $10^8 - 10^{10}$). For a thicker target, of tens of μm thick, picosecond (~ 1 – 10 ps) laser pulses are also acceptable and the contrast ratio can be much lower. It is suggested that for non-relativistic laser intensities of $I_L < 10^{18}$ W/cm^2, the TNSA mechanism works efficiently also for much longer, sub-ns laser pulses [45] but this conjecture has not been sufficiently proved yet.

The TNSA method of ion acceleration has several advantages, confirmed experimentally, in particular: (1) an excellent quality of the transverse intensity distribution of the ion (proton) beam; (2) very low transverse emittance of the beam; (3) moderate requirements for the laser beam quality (both for the transverse distribution and the contrast ratio of the beam) and intensity (both relativistic and sub-relativistic laser intensities are applicable); as a result, proton beams with energies of tens of MeV (up to 90 MeV in [46]) can be produced with currently attainable laser intensities ($\leq 10^{21}$ W/cm^2).

The main drawbacks of the TNSA scheme are: (1) a broad (quasi-Maxwellian) ion energy spectrum; (2) unfavorable scaling of the maximum ion energy E_{imax} with laser intensity: $E_{imax} \sim (I_L)^{0.5}$ [6,7] (as a result, E_{imax} is limited in practice to several hundred MeV, since I_L cannot surpass 10^{22} – 10^{23} W/cm^2 in this method); (3) relatively small areal ion density at the source $\sigma_i < 10^{17}$cm^{-2} (which results in relatively low ion beam intensity at a source that is below 10^{17} W/cm^2 for presently achievable laser intensities); (4) relatively low laser-to-ions energy conversion efficiency η (typically below a few per cent, though using the double-pulse technique $\eta \approx 15\%$ was achieved [47]); (5) high sensitivity of the ion beam parameters and η on the target rear surface quality. In spite of these drawbacks, TNSA is still the most successful and recognized method of ion (proton) acceleration, having the potential to produce ion beams of parameters desirable for numerous applications. The main achievements in the generation of ion beams using this method, and their possible applications, are described in other review papers (e.g. [6,7]).

4. Ion acceleration by the RPA mechanism

The radiation pressure acceleration (RPA) [6,7,21-26] can be considered as a special, high-intensity case of what is known as the skin-layer ponderomotive acceleration (SLPA) mechanism [16-20] of laser-driven ion acceleration. This mechanism uses strong ponderomotive forces induced at the interaction of a short laser pulse with a thin and dense preplasma layer created by the laser pulse leading edge in front of a target (Figure 3). The main part of the pulse interacts with the preplasma near the critical plasma surface, producing two opposite ponderomotive forces (see, for example, [48]. Each of these forces acts on the plasma electrons placed near the critical surface, and the electrons are pushed along the force direction and locally separated from the plasma ions. The system of the electron layer and the ion layer forms a so-called double layer (e.g. [48,49]), which is a region of non-neutral plasma where a large potential drop produces a very strong electric field (the charge-separation field) that accelerates ions. Since, in general, there are induced two opposite ponderomotive forces that push electrons near the critical surface, these forces drive (through the double-layer mechanism) two ion bunches moving in a forwards and backwards direction. At high laser intensity (> 10^{20} – 10^{21} W/cm^2) and high plasma density gradient, the forward-directed ponderomotive force (the ponderomotive or radiation pressure) clearly prevails over the backward-directed one and only a high-density forward-accelerated (along the laser beam axis) ion bunch is actually produced. This high-intensity case of SLPA is usually referred to as RPA.

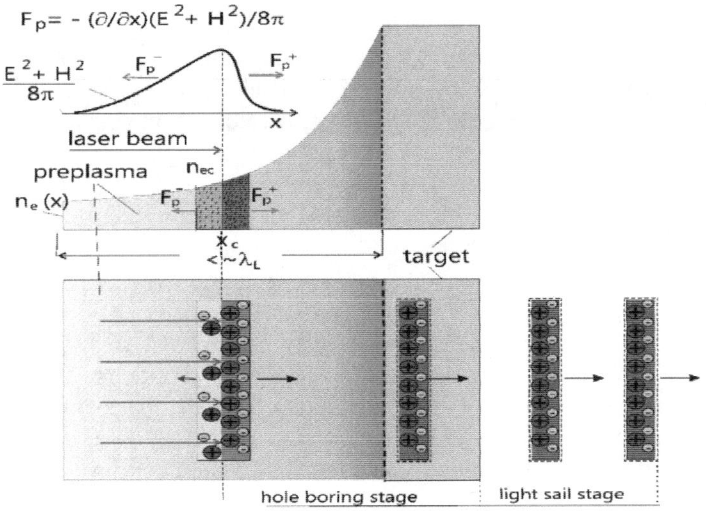

Figure 3. Simplified physical picture of ion acceleration by the RPA mechanism.

In the RPA regime of ion acceleration, two stages of acceleration can be distinguished [6,7,25,26]: the hole-boring (HB) stage and the light-sail (LS) stage (Figure 3). In the HB stage, the ions from the target front surface are accelerated towards the target interior by the moving forward charge-separation field driven by the ponderomotive (radiation) pressure. An increase in ion energies is accompanied by an increase in the number of accelerated ions and the ion bunch density since the moving double layer (in fact, its positively charged ion layer) acts as a piston on ions stored in a non-perturbed part of the target. The HB stage terminates when the accelerated ions (potentially, all ions stored in the target) reach the target rear surface. Then the LS stage begins. In this stage, the overdense plasma bunch, containing electrons and ions that cross the initial position of the target rear surface, are accelerated by the radiation pressure in free space like a sail pushed by the wind (for a more detailed description of the HB and LS stages of RPA see [6,7,25,26]). Depending on the target thickness and the laser pulse duration τ_L and fluence F_L, HB RPA or LS RPA dominates the acceleration process. When the target is sufficiently thick (or F_L is not sufficiently high) the whole laser energy is exploited for the acceleration of the ion bunch inside the target and the HB stage dominates. In turn, when the target is sufficiently thin (or F_L is sufficiently high) only a small part of laser energy is used for the formation and acceleration of the bunch inside the target (for the HB stage), and most of the laser energy is used for the acceleration of the bunch (the whole target, in practice) in the free space behind the initial target position. Thus, the LS stage dominates the acceleration process.

As mentioned in section 2, in a real ion acceleration process, two or even more acceleration mechanisms can contribute to this process, and very often the RPA mechanism competes with the TNSA mechanism in ion acceleration (sometimes RPA and TNSA produce clearly different groups of ions [19]). Since the forward-directed ponderomotive force (the radiation pressure) increases proportionally to the laser intensity, the contribution of RPA to the acceleration process grows with an increase in the laser intensity, and at very high laser intensities (well above 10^{21} W/cm^2) RPA can dominate over TNSA. This contribution also depends on the irradiated target properties as well as on the laser pulse contrast ratio (a high contrast ratio is preferred [50]) and the laser beam polarisation [6,7,22] and wavelength [51,52] (the TNSA contribution to ion acceleration can be reduced by using circular light polarisation [6,7,22] or/and a short-wavelength laser beam [51,52]).

RPA has several advantages over TNSA. The most important of them, demonstrated by numerical simulations and theory, are the following (e.g. [6,7,19,20]): (1) very high (up to tens of %) laser-to-ions energy conversion efficiency; (2) favourable scaling of ion energy with laser intensity/fluence ($E_i \sim (I\tau)^\alpha$, $\alpha = 2$ for $v_i << c$ and $\alpha = 1$ for the LS mode, or $\alpha = 1/3$ for the HB mode at $v_i \approx c$); (3) very high ion energies are possible (up to hundreds of GeV ?, limited by the laser pulse group velocity and the relativistic target transparency); (4) high areal ion density at the source, up to $\sigma \geq 10^{20}$ cm^{-2} (an ion beam of very high intensity/power/energy fluence can be produced even at moderate ion energies); (5) quasi-monoenergetic ion energy spectrum; (6) very short (down to fs) ion pulses can be produced); (7) both light and heavy ions can be efficiently accelerated.

RPA also has some drawbacks. The most significant are the following [6,7]: (1) very high laser intensities (> 10^{21} W/cm^2) are required for efficient acceleration (circularly polarised or short-wavelength laser beams are preferred); (2) the transverse homogeneity of the laser beam must be high; (3) Rayleigh-Taylor-like (and other) instabilities can occur and destroy the ion beam; (4) most RPA properties have not yet been confirmed experimentally.

RPA has been much less studied experimentally than TNSA, with investigations only performed for relatively low laser intensities from the range 10^{19} – 10^{21}W/cm^2 [6,7,19,50,53] in which ion acceleration is significantly influenced by other acceleration mechanisms (mostly TNSA). The main achievements of these studies include the measurements of ion (proton) energies of 93 MeV [50] and proton beam current densities and intensities at the source ~ 10^{12} A/cm^2 and ~ 10^{18} W/cm^2 respectively [19], as well as a rough confirmation of the quadratic scaling of the ion energy with laser intensity [50,53]. For higher laser intensities, where RPA can dominate the ion acceleration process, this mechanism has been investigated only numerically, using particle-in-cell (PIC) codes (e.g. [6,7,21-26,51,52]). Detailed experimental studies of RPA require next-generation short-pulse lasers with laser powers and intensities much higher than those of the lasers used currently.

5. Ion acceleration at ultra-relativistic laser intensities

Laser-driven ion beams have the potential to be used in many branches of science, as well as in technology and medicine. In particular, they can be applied in:
- nuclear and particle physics,
- high energy-density physics (e.g. in research on the extreme state of matter),
- inertial confinement fusion (e.g. for proton/ion fast ignition of DT fuel),
- materials science (e.g. for ion implantation),
- nuclear medicine (in hadron cancer therapy or for the production of radio-isotopes),

- radiography (e.g. for ultra-fast proton radiography).

However, the vast majority of these applications require next-generation laser drivers with a laser power or intensity higher than achievable presently. The Extreme Light Infrastructure (ELI) [54] currently being built in Europe is just a laser infrastructure where the feasibility of various potential applications of laser-driven ion beams can be tested.

5.1. Extreme Light Infrastructure

ELI will be the first laser infrastructure in the world enabling the investigation of light-matter interaction in the ultra-relativistic intensity regime (at $I_L \geq 10^{23}$ W/cm^2, where not only electrons but also protons can be accelerated to relativistic velocities) and will be a platform for both pioneering fundamental research and applications in various branches of science, technology and medicine. The primary mission of ELI consists of producing a new generation of laser-driven sources of electromagnetic radiation from the THz range to gamma-rays and bunches of electrons, protons and ions that will be used as research tools in many scientific disciplines and in the development of new technologies. ELI comprises three pillars located in three European countries, in particular: ELI-Beamlines (in the Czech Republic), ELI-Nuclear Physics (in Romania) and ELI-ALPS (in Hungary). All of them will be equipped with multi-PW femtosecond lasers. The 10 PW lasers at ELI-Beamlines (ELI-BL) and ELI-Nuclear Physics (ELI-NP) will be able to produce fs laser pulses of ultra-relativistic intensities (up to ~ $10^{23} - 10^{24}$ W/cm^2) and are planned to be used, among other things, in research on the laser-driven acceleration of ions and their various applications in plasma physics, high energy-density physics, nuclear astrophysics and in biomedical and material sciences [54]. However, studies on ion acceleration in the ultra-relativistic intensity regime (in which the generated ion beams can make many of these applications feasible) are at a very initial stage [21,55-58] and need to be intensely developed. Below, some example results of numerical investigations on the acceleration of protons (from a CH target) and super-heavy ions (from a thorium target) driven by an fs laser pulse of ultra-relativistic intensity are presented. The investigations were performed by the IPPLM team using a fully electromagnetic, relativistic 2D3V particle-in cell (PIC) code.

5.2. Acceleration of protons at ultra-relativistic laser intensities

The studies of proton acceleration by an fs laser pulse of ultra-relativistic intensity were carried out for the laser pulse parameters predicted for the ELI-BL 10 PW laser. The numerical PIC simulations were performed for the CH target with a thickness L_T ranging from 0.1 μm to 1 μm and a transverse dimension equal to 12 μm. The molecular density of the target corresponded to solid-state density and was equal to 4.86*10^{22} molecules/cm^3, and the target components (C, H) were fully ionised. A pre-plasma layer with a thickness of 0.25 μm and the density shape described by an exponential function was placed in front of the target. The laser pulse shape in time and space (along the y-axis) was described by a super-Gaussian function with a power index equal to 6. The laser beam of linear or circular polarisation had a width (FWHM) of 8 μm and the laser pulse length (FWHM) and wavelength were equal to 130 fs and 800 nm respectively. The simulations were carried out in the s, y space of dimensions 160*32 μm^2, and the number of macro-particles was assumed to be 5*10^6 for the thinner targets ($L_T \leq 0.3$ um) and 9*10^6 for the thicker targets ($L_T > 0.3$ um). Example results of the simulations are presented in Figures 4, 5 and 6.

Figure 4 presents a 2D spatial distribution of charge density of protons (a, c) and carbon ions (b, d) generated from the CH target with a thickness of 0.1 um for the final stage of ion acceleration (the simulation time t = 0.4 ps). The target was irradiated by a linearly-LP (a, b) or circularly-CP (c,d) polarised laser pulse with an intensity equal to 10^{23} W/cm^2 for CP and 2*10^{23} W/cm^2 for LP (the laser energy fluencies were approximately the same for both types of polarisation). It can be seen that for both polarisations protons are accelerated to relativistic velocities in the form of a compact, high-density proton (plasma) block, which is typical for the acceleration process dominated by RPA in the LS mode (sec. 4). For both polarisations, the proton energy spectrum is quasi-monoenergetic (slightly broader for LP), and the mean proton energy is approximately the same – about 1.8 GeV (Fig.5). In

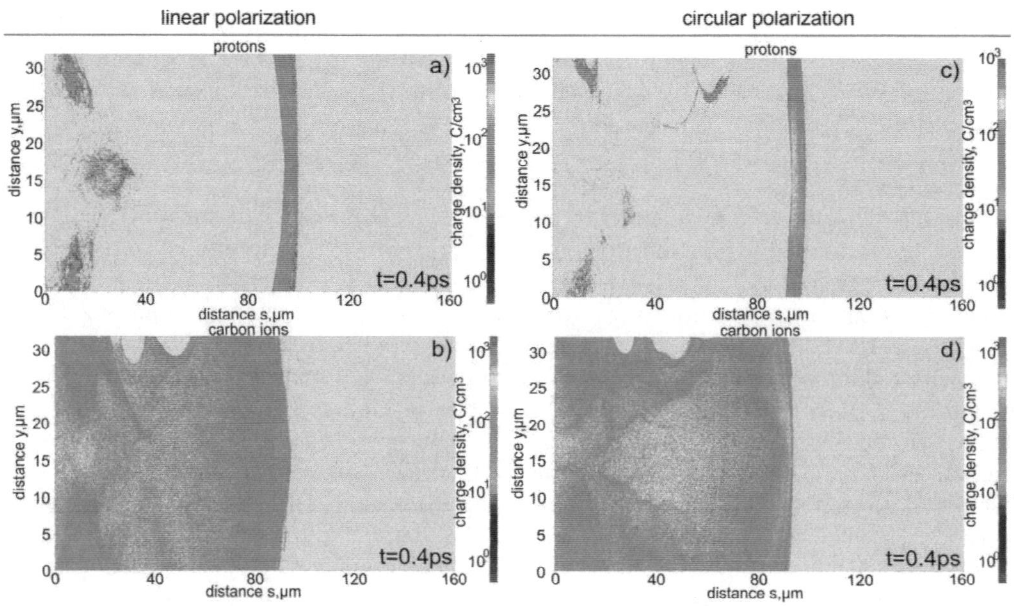

Figure 4. 2D spatial distributions of charge density of protons (a, c) and carbon ions (b, d) generated from the 0.1-μm CH target. The laser pulse intensity is equal to 10^{23} W/cm^2 for CP and $2*10^{23}$ W/cm^2 for LP.

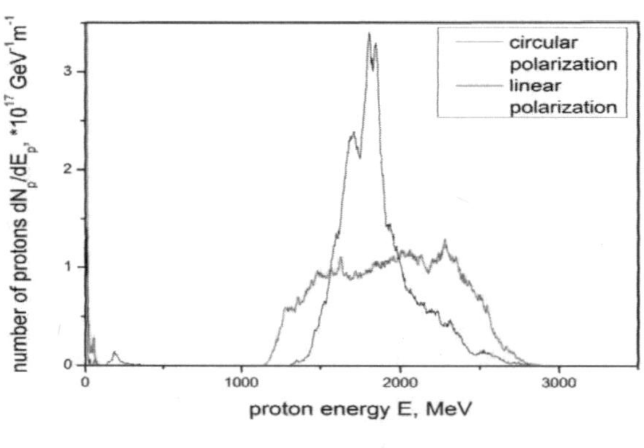

Figure 5. The energy spectra of protons generated from the 0.1-μm CH target. The laser pulse intensity is equal to 10^{23} W/cm^2 for CP and $2*10^{23}$ W/cm^2 for LP.

turn, C ions are broadly dispersed in space for both polarisations and, as a result, the carbon ion energy spectra are broad. However, the mean and maximum energies of C ions are similar for LP and CP, equal to 5.0 GeV and 19.8 GeV for LP, and 5.8 GeV and 21.0 GeV for CP respectively. A more detailed analysis shows that an important role in the broadening of the energy spectrum of carbon ions is played by the relativistic transparency of the target plasma. In the case of protons, this effect had a small influence on the properties of the accelerated proton beam, since the high-density electron-proton double layer formed in the initial stage of laser-target interaction and pushed forward by the laser field was opaque for the laser beam during the whole acceleration process.

Since the thickness l_p of the proton bunch generated from the CH target is small ($l_p < 5$μm, Fig. 4) and its velocity is very high ($v_p \approx c$), the time duration of the proton pulse τ_p (which is equivalent to the time of interaction between the bunch and a plane surface of any target perpendicular to the bunch propagation axis $t_{int} \sim l_p/v_p$) is very short (in the multi-fs range). In turn, the high density of the bunch, coupled with its high velocity, results in an extremely high intensity of the bunch. Figure 6 presents a temporal shape of the proton pulse generated from the 0.1-um CH target irradiated by an LP or CP laser beam, and recorded 40 μm behind the target. It can be seen that the shapes, durations and peak intensities I_p of the proton pulse are almost identical for both polarisations and $\tau_p \approx 11$ fs, $I_p \approx 8x10^{21}$ W/cm^2.

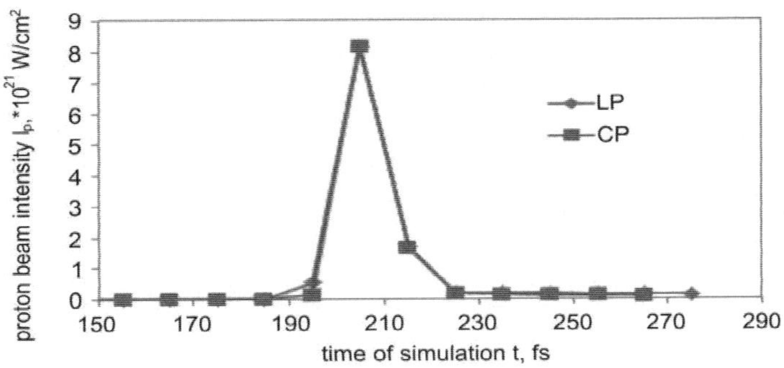

Figure 6. The temporal distribution of intensity of the proton beam generated from the 0.1-μm CH target. The laser pulse intensity is equal to 10^{23} W/cm^2 for CP and 2*10^{23} W/cm^2 for LP.

The main conclusions from the performed studies are as follows:
(1) At ultra-relativistic intensities (~ 10^{23} W/cm^2) predicted for ELI lasers, RPA is the dominant mechanism for the acceleration of protons from a sub-um plastic target, and the effect of laser polarisation on the parameters of the generated proton beam is relatively weak; for both polarisations, quasi-monoenergetic proton beams with a mean proton energy ~ 2 GeV are generated from the 0.1-μm CH target.
(2) At short distances from the irradiated target (< 50 um), the proton pulse is very short (< 20 fs), and the proton beam intensities reach extremely high values > 10^{21} W/cm^2, which are much higher than those attainable in conventional RF-driven accelerators.
(3) Such proton beams can open the door to new areas of research in high energy-density physics and nuclear physics, and can also be useful for other applications, e.g. in materials research as a tool for high-resolution proton radiography.

5.3. Acceleration of super-heavy ions at ultra-relativistic intensities

Heavy and super-heavy (A > 200) ion beams are especially important for research in nuclear and particle physics, but also find applications in other domains (e.g. in high energy-density physics). Although the possibility of a laser-driven generation of fast ions of this kind was demonstrated in experiments, both with long-pulse lasers (e.g. [59,60] and references therein) and short-pulse ones [60-65], the parameters of the produced ion beams, in particular the ion energies, were far from what was required for the above-mentioned applications. The number of numerical studies of laser-driven acceleration of ions of this kind is also very limited [66,67], and in the case of super-heavy ions numerical simulations were performed for relativistic but rather moderate laser intensities (~ 10^{20} – 10^{22} W/cm^2) [67]. In fact, it is only at ultra-relativistic intensities that the energies of super-heavy ions can reach values in the GeV or multi-GeV range required in the majority of applications. Laser-driven generation of heavy and super-heavy ions of such energies, and using them for research in nuclear physics, and particularly in nuclear astrophysics, is one of the main goals of the ELI-NP infrastructure.

One of the key research topics of the ELI-NP scientific programme is to study the production of extremely neutron-rich nuclei by a new reaction mechanism called fission-fusion, using laser-accelerated thorium (^{232}Th) ions [68]. This research is of crucial importance for understanding the nature of the creation of heavy elements in the universe; however, the proposed fission-fusion mechanism requires Th ion beams with very high ion fluencies F_i and beam intensities I_i that are unattainable in conventional RF-driven accelerators. In particular, for the ion energy per nucleon of 7 MeV/amu, the beam parameters should be as follows: $F_i \sim 10^{18}$ cm^{-2}, $I_i \sim 10^{20}$ W/cm^2 and the total number of Th ions $N_i \sim 10^{11}$ [68]. We carried out numerical simulations using our 2D3V PIC code to investigate the possibility of producing Th ion beams with the required parameters by an fs laser pulse of ultra-relativistic intensity to be generated by the ELI-NP 10 PW laser.

The results of the numerical simulations presented below were obtained for a flat thorium target with a thickness of 0.4 um and a transverse dimension equal to 12 μm. The molecular density of the target corresponded to the solid-state density and was equal to 3.04x10^{22} molecules/cm^3, where the ionization level of Th was assumed to be 30. A pre-plasma layer of 0.25 μm thickness and a density shape described by an exponential function was placed in front of the target. The target was irradiated by a circularly polarised 20 fs laser pulse with an intensity of 10^{23} W/cm^2, the beam width was 3 μm (FWHM) (the laser energy of 150 J) and the laser wavelength was 0.8 μm. The laser beam shape in time and space (along the y-axis) was described by a super-Gaussian function with a power index equal to 6.

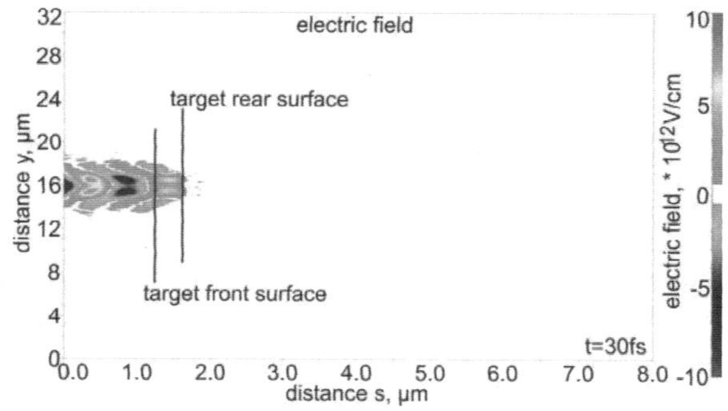

Figure 7. 2D spatial distribution of the electric field component E_y (perpendicular to the laser beam axis) at t = 30 fs when the interaction of the laser pulse with the thorium target is being terminating.

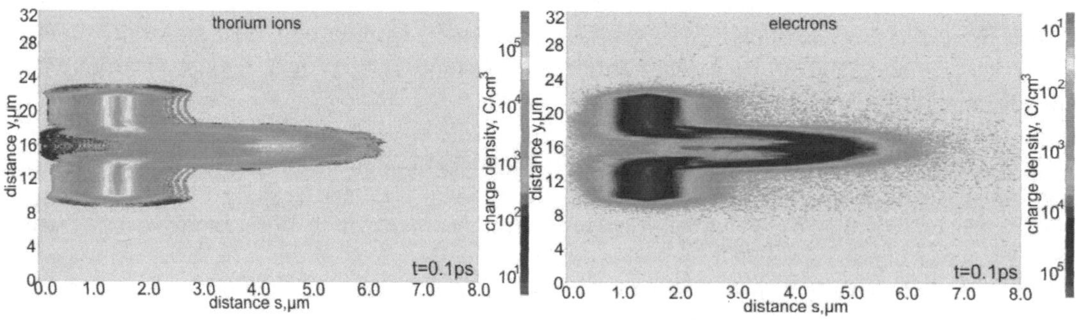

Figure 8. 2D spatial distributions of charge density of thorium ions (left) and electrons (right) for a late stage of ion acceleration (t = 100 fs).

Figure 7 presents the 2D spatial distribution of the laser electric field E_y (perpendicular to the laser beam axis) at t = 30 fs when the direct interaction of the beam with the thorium target is being terminating. The laser beam penetrates through the target, pushing forward a dense plasma bunch and at the presented stage (t = 30 fs) the laser beam reaches the target rear surface while the plasma bunch is pushed out behind this surface and starts to move into free space (not shown in the figure). This is a typical picture for HB RPA accompanied with an onset of the LS stage of acceleration (section 4). However, the LS stage of the plasma bunch acceleration cannot develop further, since the laser pulse has terminated and the ponderomotive (radiation) pressure is equal to 0. In spite of this, the ions are still accelerated during a time period much longer than the laser pulse duration. This can be seen in Figure 8, where 2D spatial distributions of the charge density of Th ions (left) and electrons (right) are shown for a late stage of acceleration (at t = 100 fs, when the acceleration time is longer than the laser pulse duration by a factor of 5). At this stage, a significant part of the plasma electrons move together with the ions and a highly collimated and relatively dense plasma (ion) beam is formed in the central

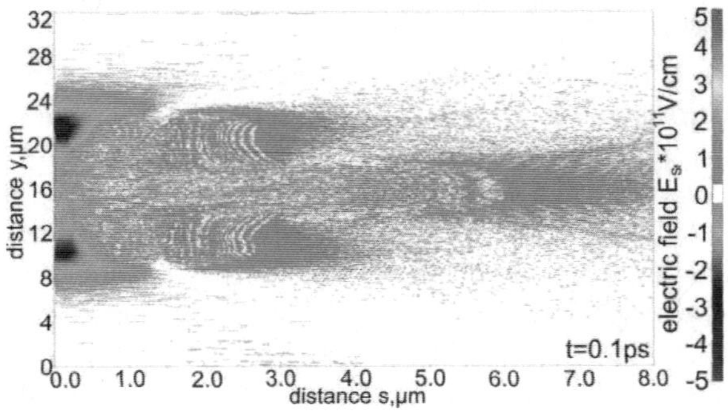

Figure 9. 2D spatial distribution of the electric field component E_s (parallel to the laser beam axis) for a late stage of ion acceleration (t = 100 fs).

part of the target. For a better understanding of the acceleration mechanism at this stage, a spatial distribution of the electric field in the plasma along the laser beam axis E_s (the field directly accelerating ions) is shown in Figure 9. It can be seen that, even at a long distance from the target initial position, the E_s field is very strong with the maxima ~ 100 GeV/cm both in front of the central ion beam and on the side parts of the target not irradiated by the laser beam. A source of this field is hot electrons (produced at the RPA stage) with energy sufficiently high to escape from the ions and to create a negative sheath (virtual cathode) able to accelerate them. This field effectively accelerates ions in the late (post-RPA) stage of the acceleration process. However, the ions from the central part of the target are much faster than those from the side parts, in spite of the fact that the accelerating fields E_s are similar for all these parts. It can be supposed that the main reason for the much higher velocities of the ions from the central part is the ballistic movement of these ions initiated in the RPA stage. Due to very high mass of the Th ions, the ballistic movement has a significant influence on the ion acceleration process in the late stage and, in particular, is a probable cause of the observed collimation of the ion beam.

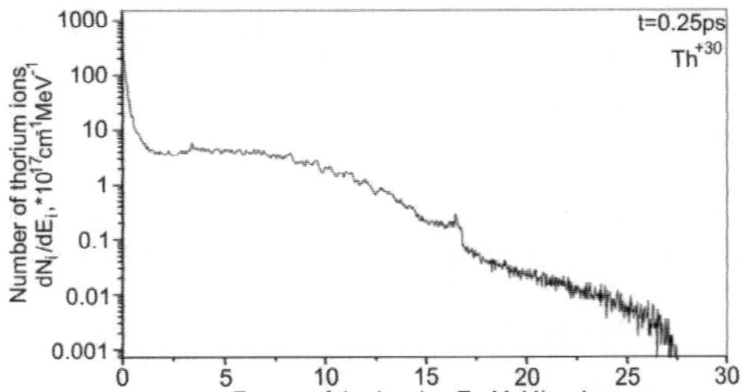

Figure 10. The energy spectrum of thorium ions at the final stage of ion acceleration (t = 250 fs).

Contrary to the case of proton acceleration considered in the previous sub-section (5.2), the accelerated ions are rather broadly dispersed along the main ion propagation direction and, as a result, the ion energy spectrum is wide – Figure 10. The main reason for this is the spatial inhomogeneity of the accelerating field at the sheath acceleration stage. In spite of the broad energy spectrum, the parameters of the produced ion beam are fairly high. For the central part of the target, with a diameter of 6 um (where the main ion beam is generated – Figure 8), these parameters are as follows: the mean ion energy $<E_i> \approx 0.9$ GeV (3.9 MeV/amu), the maximum ion energy $E_{im} \approx 6.6$ GeV (27.6 MeV/amu), the ion fluence $F_i \approx 0.9 \times 10^{18}$ cm^{-2}, the peak ion beam intensity (2 μm behind the target) $I_i \approx 1.4 \times 10^{22}$ W/cm^2, and the total ion number $N_i \approx 3.4 \times 10^{11}$.

The main conclusions from these preliminary investigations are as follows:

(1) The acceleration process of super-heavy ions at ultra-relativistic laser intensities seems to be more complex than in the case of light ions or protons; in this process, the RPA stage is followed by the sheath acceleration stage and a significant role in the second stage is played by the ballistic movement of ions accelerated at the RPA stage; this fairly complex acceleration mechanism leads to the production of a collimated high-energy ion beam of broad energy spectrum;

(2) The 20-fs laser pulse with an intensity of ~ 10^{23} W/cm^2 and an energy of 150 J is capable of producing, from the 0.4-um Th target, a thorium ion beam with more than 10^{11} ions of GeV energies (with the maximum ion energy > 6 GeV), the ion beam intensity > 10^{21} W/cm^2 and the ion fluence ~ 10^{18} ions/cm^2. The last two values are much higher than attainable in conventional accelerators, and are fairly promising for the planned ELI-NP experiments.

(3) One of the challenges that must be overcome is finding a way to narrow the energy spectrum of the accelerated ions without a dramatic reduction in other ion beam parameters, particularly in the ion beam intensity and fluence.

6. The main challenges faced by research on laser-driven ion acceleration

In spite of the considerable progress in research on laser-driven ion acceleration during the last two decades, the achieved parameters of laser-driven ion beams are still far from what is required in most of the applications mentioned in section 5. To reach these requirements, various parameters of the ion beams should be improved, and the set of parameters that would need to be upgraded depends on the type of application envisaged.

For the application of laser-driven ion beams in nuclear and particle physics, the ion energies should be considerably increased and the ion energy spectrum width would have to be decreased. A significant narrowing of this spectrum (down to 1-2 %) is also a key challenge for medical applications such as the hadron cancer therapy. Both for applications in nuclear/particle physics and medicine, the ion beams should be generated with a high repetition rate (~ 10 Hz or higher). Furthermore, since in these applications the ion beam must usually be transported over a relatively long distance (from tens of cm to tens of m), the angular divergence of the beam would have to be minimised. A promising ion acceleration mechanism with the potential to increase ion energies up to tens or even hundreds GeV is RPA working in the LS mode. However, obtaining very narrow ion energy spectrum (with a width comparable to that of conventional accelerators) with RPA seems to be an extremely difficult task. For producing such narrow spectra, it is possible that the emerging ion acceleration mechanisms like ISWA [31,32] will be more effective.

Applications of ion beams in high energy-density physics or inertial confinement fusion do not require very high ion energies (multi-GeV or higher) or a very narrow ion energy spectrum and moderate values of ion energy (from the 1 MeV – 1 GeV range) and spectrum width (~ tens %) are usually sufficient. However, the values of ion beam power/intensity or ion current/current density, as well as the ion fluencies required for these applications are very high, sometimes much higher [28,69] than those attainable in RF-driven accelerators operating currently. The ion beams of such extreme parameters can also be useful in nuclear-physics experiments for studies of nuclear reactions of very low cross sections (such as that mentioned in sec. 5). Due to the ability to produce very dense and compact ion beams (both from very thin – tens of nm and "thick" – up to tens of um), RPA is the most promising mechanism capable of producing ion beams of the parameters required in these applications. RPA-driven intense ion/proton beams can also be useful for applications in materials science, since the very short proton pulses produced by RPA can probe and diagnose material properties with unprecedented temporal and spatial resolution.

The vast majority of applications of ion beams require rather compact and economical devices that are flexible enough to be used in a variety of conditions. In the case of laser-driven accelerators, the largest and most costly element is the laser driver. Since the size, cost and complexity of the driver grows with the increase in energy of the laser pulse generated by the driver, minimising this energy without a remarkable reduction in the ion beam parameters is necessary in order to build a laser-driven accelerator useful for applications. From this point of view, the main challenge is to increase the laser-to-ions energy conversion efficiency to a level ensuring a compactness and cost of the accelerator enabling its practical usefulness. Also in this aspect, RPA seems to be the most promising method of ion acceleration, especially at ultra-relativistic laser intensities when the conversion efficiency can reach even tens of per cent.

The practical usefulness of laser-driven accelerators depends essentially on whether they can be competitive with conventional RF-driven accelerators, and whether at least some parameters of laser-driven ion beams can be comparable to or higher than those achieved in conventional accelerators. Based on the current knowledge, it is reasonable to suppose that, in the foreseeable future (say two next decades), energies of ions produced by laser-driven accelerators will still be much lower than the energies achieved currently in the largest RF-driven accelerators. Obtaining very narrow ion energy spectra and high average ion beam powers in laser-driven accelerators comparable to those attained in the conventional accelerators is also questionable. On the other hand, achieving peak ion beam powers/intensities, peak ion current/current densities or ion fluencies higher than attainable presently in the RF-driven accelerators seems to be feasible within the next decade. However, to reach this goal, next generation multi-PW laser drivers have to be used, and the ion acceleration mechanism must be better understood and controlled.

7. Summary

Laser-driven ion acceleration is a new, rapidly growing field of research, and one of the most important applications of high-peak-power lasers and laser-produced plasma.

The very high strength of laser-induced electromagnetic fields in the plasma allows the acceleration of ions up to relativistic velocities over sub-mm distances; this opens the prospect of constructing a substantially new generation of ion accelerators – more compact, less complex, and much cheaper than conventional accelerators currently operating.

Among the several laser-based methods of ion acceleration proposed so far, the methods using the TNSA mechanism and the RPA mechanism are currently most developed.

Due to its ability to accelerate ions to relativistic velocities and very high energetic efficiency, RPA seems to be the most promising ion acceleration mechanism for the use in future laser-driven accelerators; detailed experimental studies of RPA still require, however, higher laser powers and intensities than attainable currently.

Extreme Light Infrastructure will be the first laser infrastructure enabling the production of fs laser pulses of ultra-relativistic intensities and an investigation of the RPA properties, as well as testing the feasibility of various potential applications of laser-accelerated ions.

Very high intensities, powers and fluencies, and very short durations (down to tens of fs) of laser-driven ion beams generated at ultra-relativistic laser intensities provide the prospect of a variety of innovative applications of these beams in scientific research (nuclear physics, high energy-density physics, inertial fusion), as well as in technology (material engineering) and medicine (hadron cancer therapy, production of isotopes for PET). To achieve this prospect, further development of laser drivers is necessary, along with a better understanding and control of ion acceleration mechanisms.

Acknowledgments

The author acknowledges J. Domański for performing numerical simulations presented in this paper. This work was supported in part by the Eurofusion Consortium, project AWP15-ENR-01/CEA-02.

8. References

[1] Linlor W L, 1963 *Appl. Phys. Lett.* **3**, 210.
[2] Lichtman D and Ready J F, 1963 *Phys. Rev. Lett.* **10**, 342.
[3] Clark E L, Krushelnick K, Davies J R, et al. 2000 *Phys. Rev. Lett.* **84**, 670.
[4] Maksimchuk S Gu, Flippo K and Umstadter D, 2000 *Phys. Rev. Lett.* **84**, 4108.
[5] Snavely R A, Key M H, Hatchett S P, et al. 2000 *Phys. Rev. Lett.* **85**, 2945.
[6] Macchi A, Borghesi M and Passoni M, 2013 *Rev. Mod. Phys.* **85**, 751, and references therein.
[7] Daido H, Nishiuchi M and Pirozhkov A S, 2012 *Rep. Prog. Phys.* **75**, 056401, and references therein.
[8] Wilks S C, Langdon A B, Cowan T E, et al. 2001 *Phys. Plasmas*, **8**, 542.
[9] Mackinnon A J, Borghesi M, Hatchett S, et al. 2001 *Phys. Rev. Lett.* **86**, 1769.
[10] Badziak J, Woryna E, Parys P, et al. 2001 *Phys. Rev. Lett.* **87**, 215001.
[11] Mackinnon A J, Sentoku Y, Patel P K, et al. 2002 *Phys. Rev. Lett.* **88**, 215006.
[12] Borghesi M, Mackinnon A J, Campbell D H, et al. 2004 *Phys. Rev. Lett.* **92**, 055003.
[13] Cowan T E, Fuchs J, Ruhl H, et al. 2004 *Phys. Rev. Lett.* **92**, 204801.
[14] Allen M, Patel P K, Mackinnon A, et al. 2004 *Phys. Rev. Lett.* **93**, 265004.
[15] Borghesi M, Fuchs J, Bulanov S V, et al. 2006 *Fusion Sci. Technol.* **49**, 412
[16] Badziak J, Głowacz S, Jabłoński S, et al. 2005 *Laser Part. Beams* **23**, 143.
[17] Badziak J, Jabłoński S and Głowacz S, 2006 *Appl. Phys. Lett.* **89**, 061504.
[18] Badziak J, Głowacz S, Jabłoński S, et al. 2006 *Laser Part. Beams* **24**, 249.
[19] Badziak J, Jabłoński S, Parys P, et al. 2008 *J. Appl. Phys.* **104**, 063310.
[20] Badziak J, Mishra G, Gupta N K, Holkundkar A R, 2011 *Phys. Plasmas,* **18**, 053108.
[21] Esirkepov T, Borghesi M, Bulanov S V, et al. 2004 *Phys. Rev. Lett.* **2004,** 92, 175003.
[22] Macchi A, Cattani F, Liseykina T V, Cornalti F, 2005 *Phys. Rev. Lett.* **94**, 165003.
[23] Robinson A P L, Zepf M, Kar Set al. 2008 *New J. Phys.* **10**, 033034.
[24] Liseykina T V, Borghesi M, Macchi A, Tuveri S, 2008 *Plasma Phys. Control. Fusion,* **50,** 124033.
[25] Klimo O, Psikal J, Limpouch J, Tikhonchuk V T 2008 *Phys. Rev. ST Accel. Beams* **11**, 031301.
[26] Grech M, Skupin S, Diaw A, et al. 2011 *New J. Phys.* **13**, 123003.
[27] Yin L, Albright B J, Hegelich B M et al. 2007 *Phys. Plasmas,* **14**, 056706.
[28] Fernandez J C, Albright B J, Beg F N, et al. 2014 *Nucl. Fusion,* **54**, 054006.
[29] Denavit J. 1992 *Phys. Rev. Lett.* **69**, 3052.
[30] Silva L O, Marti M, Davies J R and Fonseca R A 2004 *Phys. Rev. Lett.* **92**, 015002.
[31] Jung D, Yin L, Albright B J, Gautier D C et al. 2011 *Phys. Rev. Lett.* **107**, 115002.
[32] Yin L, Albright B J, Jung D, Bowers K J, et al, 2011 *Phys. Plasmas,* **18**, 053103.
[33] Esirkepov T Z, Bingham R, Bulanov S V et al. 2000 *Laser Part. Beams* **18**, 503.
[34] Bychenkov V Yu and Kovaliev V E 2005 *Plasma Phys. Rep.* **31**, 178.
[35] Atzeni S and Meyer-ter-Vehn J, 2004 *The Physics of Inertial Fusion* (Clarendon, Oxford).
[36] Obenschain S P, Whitlock R R, McLean E A, et al. 1983 *Phys. Rev. Lett.* **50**, 44.
[37] Karasik M, Weaver J L, Aglitskiy Y, et al. 2010 *Phys. Plasmas* **17**, 056317.
[38] Min Shui, Genbai Chu, Bin Zhu, et al. 2016 *J. Appl. Phys.* **119**, 035903.
[39] Badziak J, Borodziuk S, Pisarczyk T, et al. 2010 *Appl. Phys. Lett.* **96**, 251502.
[40] Badziak J, Jabłoński S, Pisarczyk T, et al. 2012 *Phys. Plasmas*, **19**, 053105.
[41] Badziak J, Rosiński M, Jabłoński S, et al. 2015 *Plasma Phys. Control. Fusion* **57**, 014007.
[42] Torrisi L, Gamino S, Ando L, et al. 2006 *J. Appl. Phys.* **99**, 083301.
[43] Torrisi L, Cutroneo M, Ando L and Ullschmied J, 2013 *Phys. Plasmas* **20**, 023106.
[44] Roth M, Blazevic A, Geissel M et al. 2002 *Phys. Rev. ST Accel. Beams*, **5**, 0613001.
[45] Torrisi L, 2014 Laser Part. Beams 32, 383.
[46] Wagner F, Brabetz C, Deppert O, et al. 2016 *High Power Laser Sci. Eng.* **4**, e45.

[47] Brenner C M, Robinson A P L, Markey K, et. al. 2014 *Appl. Phys. Lett.* **104**, 081123.
[48] Hora H, *Physics of Laser Driven Plasmas (*Wiley, New York, 1981).
[49] Eliezer S, Nissim N, Martinez Val J M, et al. 2014 *Laser Part. Beams*, **32**, 211.
[50] Jong Kim I, Hong Pae Ki, Woo Choi Il, et al. 2016 *Phys. Plasmas* **23**, 070701.
[51] Badziak J, Jabłoński S, 2010 *Phys. Plasmas*, **17**, 073106.
[52] Badziak J, Jabłoński S, 2010 *Appl. Phys. Lett.* **99**, 071502.
[53] Henig A, Steinke S, Schnürer M, et al.2009 *Phys. Rev. Lett.***103**, 245003.
[54] www.eli-laser.eu
[55] Yin L, Albright B J, Hegelich B M and Fernandez J C, 2006 *Laser Part. Beams* **24**, 291.
[56] Zheng F L, Wang H Y, Yan X Q, et al. 2012 *Phys Plasmas* **19**, 023111.
[57] Xu Y, Wang J, Qi X, et al. 2017 *Phys. Plasmas* **24**, 033108.
[58] Domański J, Badziak J, Jabłoński S, 2017 *Laser Part. Beams* **35**, 286.
[59] Laska L, Jungwirth K, Kralikowa B, et al. 2003 *Plasma Phys. Control. Fusion* **45**, 585.
[60] Badziak J, 2007 *Opto-Electron. Rev.* **15**, 1.
[61] Clark E L, Krushelnick K, Zepf M, et al. 2000 *Phys. Rev. Lett.* **85**, 1654.
[62] Badziak J, Parys P, Vankov A B, et al., 2001 *Appl. Phys. Lett.* **79**, 21.
[63] Badziak J, Makowski J, Parys P, et al. 2001 *J. Phys. D: Appl. Phys.* **34**, 1885.
[64] Badziak J, Hora H, Woryna E, et al. 2003 *Phys. Lett. A* **315**, 452.
[65] McKenna P, Ledingham K W D, Yang J M, et al. 2004 *Phys. Rev. E* **70**, 036405.
[66] Wu D, Qiao B, McGuffey C, et al. 2014 *Phys. Plasmas* **21**, 123111.
[67] Petrov G M, McGuffey C, Thomas A G R, et al. 2016 *Phys. Plasmas* **23**, 063108.
[68] Negoita F, Roth M, Thirolf P G, et al. 2016 *Romanian Rep. Phys.* **68**, S37.
[69] Badziak J, Jabłoński S, and Wołowski J, 2007 *Plasma Phys. Control. Fusion* **49**, B651.

Studies of runaway electrons via Cherenkov effect in tokamaks

J Zebrowski[1], L Jakubowski[1], M Rabinski[1], M J Sadowski[1], M J Jakubowski[1], R Kwiatkowski[1], K Malinowski[1], R Mirowski[1], J Mlynar[2], O Ficker[2], V Weinzettl[2], F Causa[3,4], COMPASS[2] and FTU[3] Teams
[1] National Centre for Nuclear Research (NCBJ), 7 A. Soltana Str., 05-400 Otwock, Poland
[2] Institute of Plasma Physics of the CAS, Za Slovankou 3, 182 00 Prague 8, Czech Republic
[3] ENEA, C.R. Frascati, Via E. Fermi 45, 00044 Frascati (Roma), Italy
[4] Present affiliation: IFP-CNR, Istituto di Fisica del Plasma, Via R. Cozzi 53, 20125 Milano, Italy

Corresponding author e-mail address: Jaroslaw.Zebrowski@ncbj.gov.pl

Abstract. The paper concerns measurements of runaway electrons (REs) which are generated during discharges in tokamaks. The control of REs is an important task in experimental studies within the ITER-physics program. The NCBJ team proposed to study REs by means of Cherenkov-type detectors several years ago. The Cherenkov radiation, induced by REs in appropriate radiators, makes it possible to identify fast electron beams and to determine their spatial- and temporal-characteristics. The results of recent experimental studies of REs, performed in two tokamaks - COMPASS in Prague and FTU in Frascati, are summarized and discussed in this paper. Examples of the electron-induced signals, as recorded at different experimental conditions and scenarios, are presented. Measurements performed with a three-channel Cherenkov-probe in COMPASS showed that the first fast electron peaks can be observed already during the current ramp-up phase. A strong dependence of RE-signals on the radial position of the Cherenkov probe was observed. The most distinct electron peaks were recorded during the plasma disruption. The Cherenkov signals confirmed the appearance of post-disruptive RE beams in circular-plasma discharges with massive Ar–puffing. During experiments at FTU a clear correlation between the Cherenkov detector signals and the rotation of magnetic islands was identified.

1. Introduction

Runaway electrons (REs) are generated in tokamaks when the friction of electron collisions with plasma does not compensate the externally induced electrical force coming from the toroidal electric field E_{tor}. Such beams may reach energies up to several MeV. Therefore, some electrons may be accelerated and run away in the phase space, and finally escape from the magnetic trap. There are three main mechanisms of the runaways generation: i) Dreicer (primary) mechanism; ii) avalanche (secondary) mechanism, and iii) hot-tail mechanism. A so-called critical field is usually defined as a theoretical limit for the electrical field E_{tor}, below which RE cannot be produced by these mechanisms.

The understanding of processes of the runaway electrons emission in tokamak-type devices is of importance because of a risk of serious damages of in-vessel components, which can ultimately lead to a shut-down of the facility. Such a risk makes very difficult to perform detailed experimental studies of the RE emission in tokamaks. In fact REs can be recorded in small- and medium-size tokamaks either during the current ramp-up or the flat-top phase [1, 2] when the electron density n_e is low and/or

IOP Conf. Series: Journal of Physics: Conf. Series **959** (2018) 012002 doi:10.1088/1742-6596/959/1/012002

toroidal electric field E_{tor} is higher than the critical field. In general, various observations noted that the spontaneous emission of the post-disruptive RE is not possible if B_{tor} is below 2T [3]. However, the post-disruptive REs were achieved with B_{tor} lower than 2T in discharges with the high-Z massive gas-, or high-Z pellet-injection [4].

The NCBJ (former IPJ) team proposed to study RE by means of Cherenkov-type detectors several years ago. The Cherenkov radiation, as induced by RE in appropriate radiators and recorded with photomultipliers, enables the identification of fast electron beams as well as the determination of their spatial- and temporal-characteristics to be performed [5]. It can be treated as a novel diagnostic technique which is very convenient for direct measurements of the escaping fast REs in different tokamak facilities [6-10]. In a Cherenkov-type probe, as designed especially for multi-channel measurements, several separate channels with different low-energy detection thresholds were applied. Those thresholds were determined by materials and thicknesses of the applied absorption filters. It made possible to estimate roughly some energetic characteristics of the investigated RE beams [11]. Integrals of the Cherenkov probe signals were proportional to the recorded local electron-fluencies.

This paper presents results of the RE-emission studies performed by means of the Cherenkov-type probes in two tokamaks: COMPASS in Prague and FTU in Frascati.

2. Three-channel Cherenkov probe measurements in COMPASS

The COMPASS tokamak [12] is a small-size device with the ITER-like plasma cross-section, major radius R_0 = 0.56 m, and minor radius a = 0.23 m. Its main advantage is large flexibility and low operational costs. It can be operated both in the limited- and diverted plasma-configuration. During RE campaigns the discharge parameters were as follows: elongation k = (1–1.4), plasma current I_p = (130–180) kA, line-averaged electron density: $n_e = (1–5) \times 10^{19}$ m^{-3}, and toroidal magnetic field B_T = (1.15-1.2) T. All the mentioned examples show limited plasmas with a short elongation phase (k<1.2) before reaching a current plateau to improve the discharge stability of scenarios with circular plasmas. The experiments were usually run with pure deuterium (with Glow Discharge Cleaning in He between the shots); the typical pulse length was about 0.4 s. The population of REs was observed at densities $n_e < 4 \times 10^{19}$ m^{-3}, with a strong dependence on the gas fueling scenario and plasma shape [13].

COMPASS has a relatively low plasma current and the REs cannot achieve very high energies and fluencies. Hence, severe damages of the vessel are rather improbable [14]. It makes the COMPASS device a suitable machine for RE studies. In the recent RE-dedicated experiments argon-triggered plasma disruptions were realized and post-disruption RE beams were successfully induced there (both in current ramp-up as well as in current plateau phase) [4, 15]. To monitor escaping REs three-channel Cherenkov probe, which was equipped with three CVD-type diamond radiators with the cylindrical-tablet shapes of 8 mm-in-diameter and 1.5 mm-in-thickness, was used and located in the scrape-off plasma. The front surfaces of the mentioned radiators were covered by metal layers of different thicknesses, and it made possible to diversify low-energy thresholds for the electrons recorded by the individual channels of the probe.

In the Cherenkov probe applied for measurements at COMPASS the front surface of the radiator in Channel No 1 was covered by a Ti/Pt/Au interlayer of 1.3 μm in the total thickness, in Channel No 2 - by the same interlayer and a 35-μm-thick Mo-layer, and in Channel No 3 - by such an interlayer and a 77-μm-thick Mo-layer. Due to different materials and thicknesses of these absorption filters low-energy detection-thresholds for electrons impinging upon the described radiators were as follows: 58 keV in Channel No 1, 145 keV in Channel No 2, and 221 keV in Channel No 3. The rough calibration of the first channel of the described probe was performed at the NCBJ laboratory by means of a 100-keV electron beam obtained from an accelerator.

The most important result of the recent RE campaigns at COMPASS, from the point of view of potential applications of Cherenkov-type detectors, was the possibility to obtain the confirmation of a dependence of RE-signals on the radial position of the Cherenkov probe measuring head for circular plasma produced by deuterium discharges, as shown in figures 1(a) and 1(b).

IOP Conf. Series: Journal of Physics: Conf. Series **959** (2018) 012002 doi:10.1088/1742-6596/959/1/012002

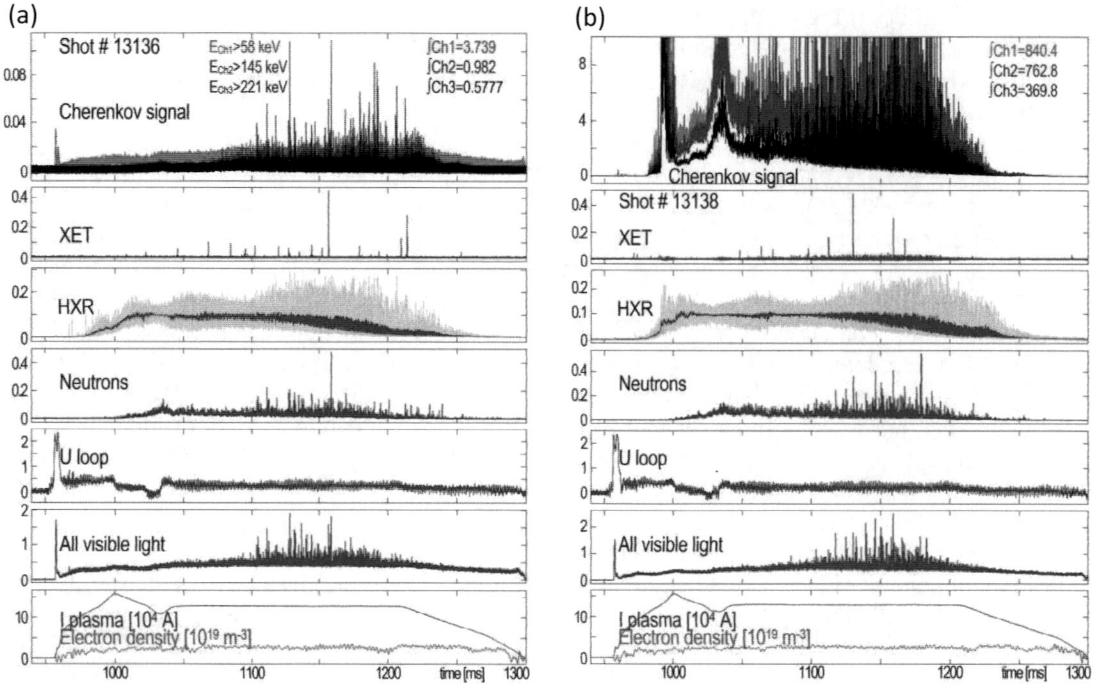

Figure 1. Comparison of electron-induced signals (obtained from the 3-channel Cherenkov probe) with the basic plasma signals: hard X-ray signal (HXR), photo-neutrons, U_{loop}, visible light emission, plasma current and line-averaged electron density for COMPASS shots with deuterium fuelling and the circular plasma. XET signal presents X-ray pulses recorded inside the shielded box of Cherenkov probe photo-multipliers. Signals intensities are shown as a function of time. Left part (a) shows signals recorded during shot #13136, with the measuring head at the position R = 0.785m (5 mm inside the tokamak vessel, 45 mm outside the midplane separatrix radius). Right part (b) shows signals recorded during shot #13138, with the measuring head at the position R = 0.765m (25 mm inside the tokamak vessel, 25 mm outside the midplane separatrix radius).

The changes of the measuring head position were performed by means of an appropriate manipulator on a shot-to-shot basis. An increase in integrals of the Cherenkov probe signals, taken for whole time of discharges, as observed for all the probe channels and presented in figures 1a and 1b, which corresponded to the increase in electron beams fluencies, was higher than two orders of magnitude.

The result of the probe position scanning, i.e. the determination of a radial dependence of RE beams fluency (obtained by the probe signals integration over the discharge time) on the radial position of the measuring head, is shown in figure 2. The presented results were taken from a series of subsequent COMPASS shots with the deuterium fuelling and the circular plasma at conditions similar to those which were applied earlier, for discharges presented in figure 1. Five radial positions of the measuring head was taken into account.

The measurements carried out by means of the three-channel Cherenkov probe showed also that the post-disruption RE beams were intensively produced in plasma discharges performed with the gas-puffing. The experiments dedicated to the RE control and mitigation at COMPASS [15-17], which were carried out with the argon-puffing, shoved that post-disruption RE beams can occur during the current ramp-up phase, and proved that the first fast-electron peak appears usually during the current quench phase, even before the hard X-rays (HXR) pulse. However, the most intense and distinct peaks

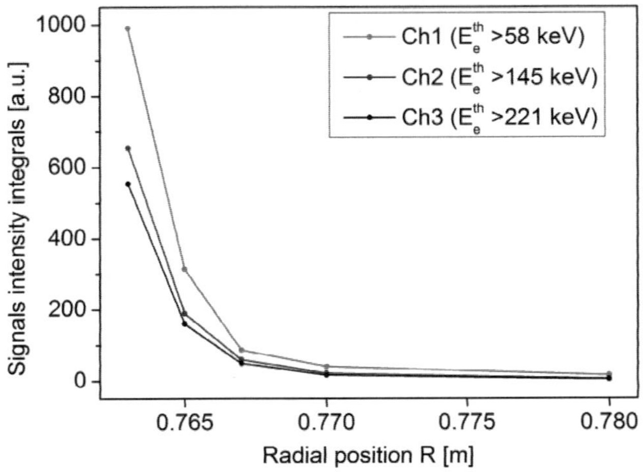

Figure 2. Dependences of the RE beams fluencies (recorded by means of the Cherenkov probe) on the radial position of the measuring head. The experimental data were taken from several COMPASS shots (from #14558 to #14563, but without #14561), which were performed with the deuterium fuelling and circular plasma.

corresponding to RE losses, in all the channels of the Cherenkov probe, appear mainly during the emission of post-disruption RE beams.

The RE confinement was better in the divertor-type configuration and in the elongated limited plasma experiment, as compared to circular limited plasma [15]. The initiation of the relatively fast decay of the RE emission in the case of circular plasma, when density increased to $n_e = 3 \times 10^{19}$ m^{-3}, was also observed. In the RE dominated discharges the maximum energy value of the REs might be estimated on the basis of the RE-generated synchrotron radiation, which could be directly measured by means of an infrared camera [17]. That estimation gave the energy value equal to about 25 MeV.

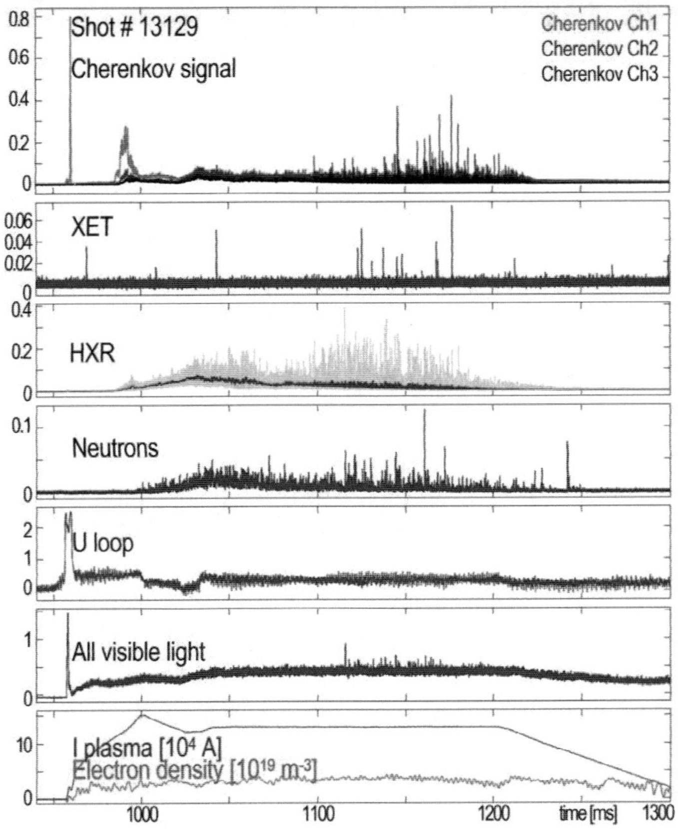

Figure 3. Comparison of time-resolved electron-induced signals (obtained from the 3-channel Cherenkov probe) with the basic plasma signals: hard X-ray signal (HXR), photo-neutrons, U_{loop}, visible light emission, plasma current and line-averaged electron density for COMPASS #13129 shot performed at the deuterium fuelling and circular plasma. The measuring head position was R = 0.775 m (15 mm inside the tokamak vessel). The line-averaged electron density was equal to 4.4 x 10^{19} m^{-3}. The front part of the Ch1 signal is a good example of a precursor character of relatively low-energy RE beams appearance in relation to the total REs and HXR emission.

It was observed that the RE beams of energy 60-140 keV, which were recorded only in channel No 1 of the Cherenkov probe for discharges with the deuterium circular plasma configuration, appeared as narrow spikes at the instant of the plasma discharge start, and at the maximum of the plasma discharge current, as can be seen in figure 3.

It should be noted that in the later phase of the plasma discharge, which was performed with the argon puffing and the circular plasma shape, the primary signal peaks recorded by means of channel No 1 (measuring electrons of energy > 58 keV) could also be treated as a precursor HXR signals which is due to high-energy RE losses [18].

During the COMPASS plasma experiments it was noticed that the appearance of magnetic islands changed completely dynamics of the RE losses in a comparison with that observed for MHD-quiescent plasma discharges. Periodic RE losses to the limiter, which are connected with the magnetic island rotation, were observed, and it was concluded that the maximum of such losses occurred when the O point of the island passed the limiter edge [16]. It was observed particularly during the last phase of the flat-top of the low-density discharges. The link between localized Cherenkov signals and MHD perturbations seemed to be a result of the plasma activity observed during disruptions caused by the massive gas injections, and particularly during those accompanied by the RE beams generation. In some discharges with magnetic islands the amplitude of RE losses seemed to be considerably lower, i.e., the magnetic island could probably confine some REs [16].

An example of multiple RE-spikes, as recorded by means of the three-channel Cherenkov probe, which usually appear during MHD instabilities presence, is shown in figure 4.

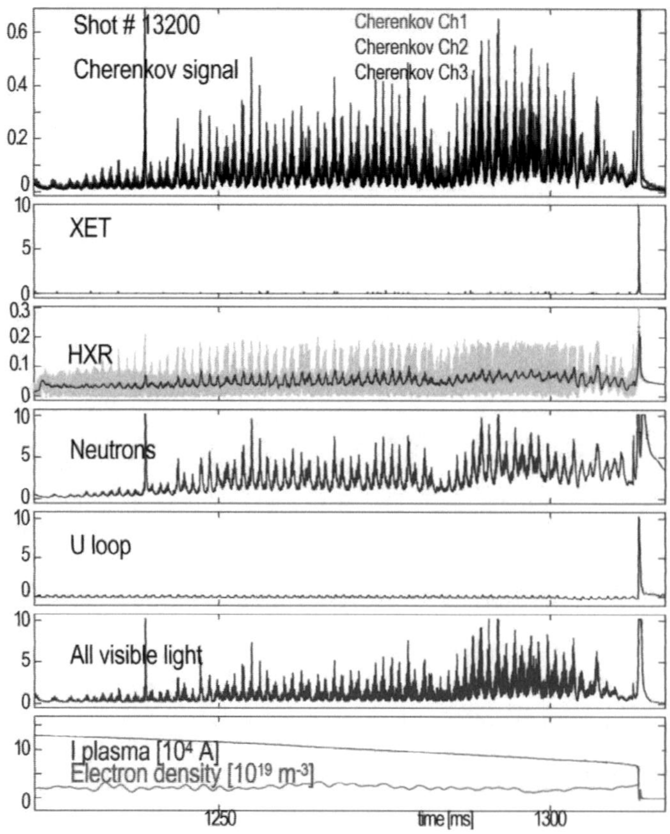

Figure 4. Comparison of electron-induced signals (obtained from the 3-channel Cherenkov probe) with the main COMPASS signals: hard X-ray signal (HXR), photo-neutrons, U_{loop}, visible light emission, plasma current and plasma density for the final part of COMPASS #13200 shot performed with the deuterium filling and circular plasma. XET signal presents X-ray pulses recorded inside the shielded box of Cherenkov probe photo-multipliers. Signals intensities are shown as a function of time. The measuring head position was R = 0.790 m (at the level of the tokamak vessel wall). The line-averaged electron density was equal to 3.5 x 10^{19} m^{-3}. Multiple RE spikes, which are well recorded by means of the Cherenkov probe, appear usually during MHD instabilities.

3. Single-channel Cherenkov probe measurements in FTU

The FTU tokamak [20] is a compact high magnetic-field device with the circular poloidal cross-section, major radius R_0 = 0.935 m, and minor radius a = 0.30 m. The FTU vacuum chamber is made of a stainless-steel. There is a toroidal limiter made of molybdenum tiles at the high field side, and additionally an outer molybdenum poloidal limiter (at the low field side), and a vertical poloidal lithium-limiter. The tokamak parameters during RE studies were as follows: toroidal magnetic field B_T from 2 T to 8 T, plasma current I_p from 0.2 MA to 1.6 MA.

For measurements of RE beams escaping from plasma in the FTU tokamak the use was made of a dedicated one-channel Cherenkov probe designed by the NCBJ team. A titanium-zirconium-molybdenum (TZM) alloy was applied to manufacture the measuring-head body. That head was equipped with a CVD-type diamond radiator of the cylindrical-tablet shape of 10 mm in diameter and 1 mm in thickness. The front surface of the radiator was coated with a Ti/Pt/Au layer of 1.3 μm in the

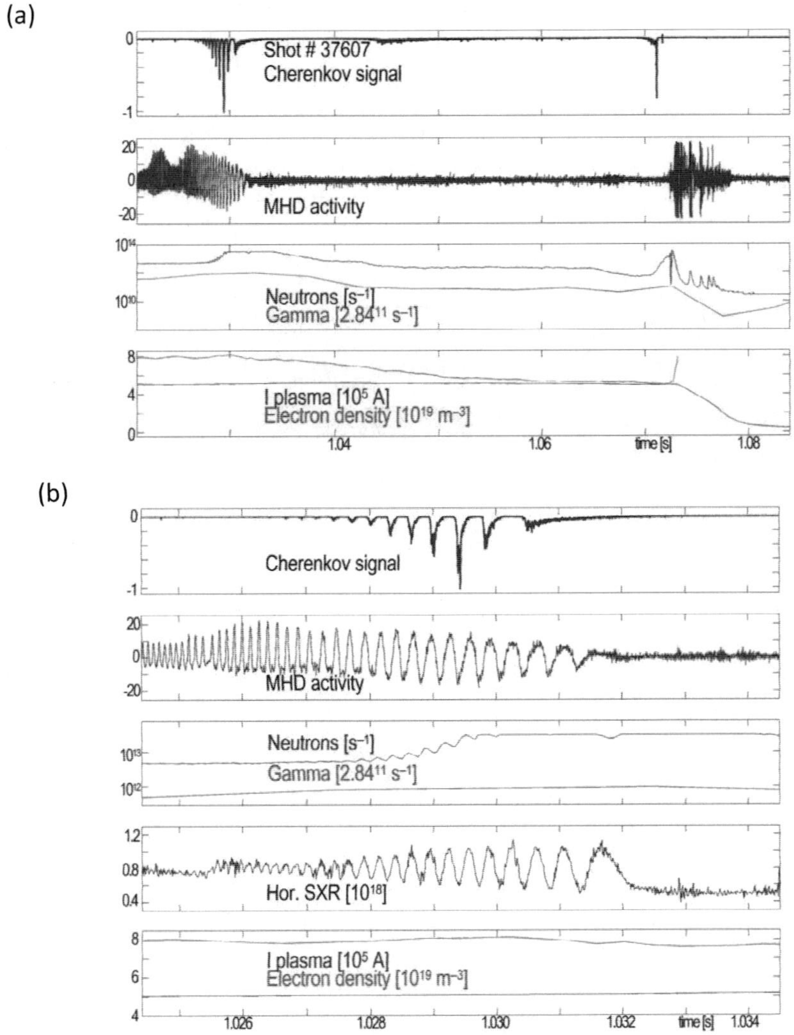

Figure 5. Comparison of the Cherenkov-probe signals with the MHD activity signals (recorded by means of a Mirnov coil), as well as signals from neutron- and gamma- detectors, and waveforms showing the plasma current amplitude and line-averaged electron density for the FTU shot #37607, performed at B_T = 5.3 T and I_p = 500 kA: (a) – general view of the observed correlations, (b) – a period of the discharge (zoomed up), when the rotation of magnetic islands was identified.

total thickness. The low-energy detection threshold for electrons impinging upon the radiator was 58 keV. The Cherenkov radiation from this radiator was transferred, in the first section, through a polished metal tube and then it was coupled to a high-gain photomultiplier through a VIS/UV optical fiber.

During the reported experiments the Cherenkov probe was inserted into the FTU tokamak vessel at variable positions in a shadow of the limiter. The detector was not sensitive to the background electromagnetic emission, i.e. the visible-, synchrotron- and gamma- radiation.

Cherenkov signals were observed not only during discharge disruption, but also in correlation with MHD instabilities [21, 22].

Signal modulation was due to the passage of the magnetic island in front of the probe, as demonstrated by the clear phase-relation between the Cherenkov signals, the ECE signals, and those from a proton-recoil NE213 scintillator detector. The Cherenkov signal vanished when the O-point moved away from the probe, and it reached maximum when the O-point of the island is in front of the probe [22]. The last rotation of the magnetic island before disruption was clearly detected by the Cherenkov probe as a relatively large signal.

An example of close correlations of signals obtained from the Cherenkov probe, Mirnov coils (MHD activity), neutron- and gamma-cameras, and soft x-ray cameras, as well as signals describing of plasma current intensity and linear-averaged plasma density, is shown above in figure 5. It should be mentioned that in the FTU shot #37607 (analyzed in this figure) some amount of neon was puffed in the vessel at the instant $t = 0.5$ s for 50 ms in order to seed MHD activity.

The reported studies provided an evidence of a loss of the confinement of fast electrons, which was driven by high-amplitude magnetic islands. In addition, as discussed elsewhere [22, 23] The Cherenkov-probe signals can prove that an additional RE expulsion mechanism is the magnetic perturbation due to amplitude fluctuations of a magnetic island.

4. Conclusions

During the recent RE campaigns at COMPASS the fast electron-beams measurements, which were performed by means of the three-channel Cherenkov probe, confirmed good capabilities of such a probe, particularly for the carrying out an analysis and drawing conclusions concerning energetic characteristics of RE beams. The Cherenkov-type probe has also proved experimentally a strong dependence of the RE signals on the radial position of the measuring head.

The RE dynamics in the presence of magnetic islands at COMPASS could also be studied by means of multi-channel Cherenkov probe. Research on the correlation of the localized Cherenkov signals with MHD perturbations could provide insight into the analysis of the influence of the magnetic island on RE losses. It should be noted that the Cherenkov probe was able to detect prompt losses of low energy RE beams, which accompanied the current quench phenomena. It was also stated that the RE losses (measured by means of the Cherenkov probe or HXR detector), as well as bursts of magnetic perturbations and fast-particle-related signals (recorded by the ECE diagnostics), were well correlated with images of plasma filaments (observed by the fast visible-radiation Phatron camera) [18].

The observations of correlations of signals from the Cherenkov-probe with those obtained from the other diagnostics, including signals of the primary MHD activity in the FTU tokamak, revealed significant opportunities for the use of the Cherenkov-type diagnostics in order to widen the knowledge of RE losses physics, particularly in scenarios involving plasma instabilities which can lead to disruptions.

In both tokamaks, i.e in COMPASS and FTU, fast electron losses could be effectively monitored by means of diagnostic technique based on Cherenkov effect occurring in the appropriate radiators located inside the tokamak vessels. It made Cherenkov probes potentially interesting for applications in the prediction and control of RE phenomena associated mainly with strong MHD activity in tokamaks.

Acknowledgments
This work has been carried out within the framework of the EUROfusion Consortium and has received funding from the Euratom research and training programme 2014-2018 under grant agreement No 633053. The views and opinions expressed herein do not necessarily reflect those of the European Commission. The reported scientific studies were partly supported by the Polish Ministry of Science and Higher Education within a framework of the scientific financial resources in the years 2016-2017, allocated for the realization of the international co-financed projects. The experiments on COMPASS were supported by the MEYS CR project No. LM2015045.

References

[1] Jaspers R J E 1995 *Relativistic runaway electrons in tokamak plasmas* PhD Thesis Technical University, Eindhoven (158pp)
[2] Esposito B, Martin-Solis J R, Poli F M, Mier J A, Sanchez R and Panaccione L 2003 *Physics of Plasmas* **10** (6) 2350-23
[3] Gill R D, Alper B. de Baar M, Hender T C, Johnson M F, Riccardo V and contributors to the EFDA-JET Workprogramme 2002 *Nuclear Fusion* **42** (8) 1039-1044
[4] Vlainic, M. Mlynar J, Cavalier J, Weinzettl V, Paprok R, Imrisek M, Ficker O, Varavin, Vondracek M P, Noterdaeme J-M and the COMPASS Team 2015 *J. Plasma Phys.* **81** 475810506
[5] Sadowski M J, Jakubowski L and Szydlowski A 2004, *Czech. J. Phys.* **54**, C74-C80
[6] Jakubowski L, Stanisławski J, Sadowski M J, Zebrowski J, Weinzettl V and Stockel J 2006 *Czech. J. Phys.* **56 Suppl. B** B98-B103
[7] Zebrowski J, Jakubowski L, Sadowski M J, Malinowski K, Jakubowski M J, Weinzettl V, Stockel J, Vacha M and Peterka M 2008 *AIP Conf. Proc.* **993** 255-258
[8] Jakubowski L, et al. *Rev. Sci. Instrum.* **81** (2010) 013504 (9pp)
[9] Jakubowski L, Malinowski K, Sadowski M J, Zebrowski J, Plyusnin V V, Rabinski M, Fernandes H, Silva C, Duarte P and Jakubowski M J 2010 *Nucl. Instr. Meth. A* **623** 686-689.
[10] Jakubowski L, et al. 2013 *Rev. Sci. Instrum.* **84** 016107 (3pp)
[11] Jakubowski L, Plyusnin V V, Sadowski M J, Zebrowski J, Malinowski K, Rabinski M, Fernandes H, Silva C, Duarte P and Jakubowski M 2012 *Nukleonika* **57** (2) 177-181
[12] Panek R, et al. 2016 *Plasma Phys. Control. Fusion* **58** 014015 (9pp)
[13] J. Mlynar J, et al. 2016 EUROFUSION **WPMST1-CP(16)** 15152 - submitted for publication in Proc. 26th IAEA Fusion Energy Conference (8pp)
[14] Vlainic M, Mlynar J, Weinzettl V, Paprok R, Imríšek M, Ficner O, Vondracek P and Havlicek J 2015 *Nukleonika* **60** (2) 249-255
[15] Mlynar J, et al. 2015 *in Proc. 42nd EPS Conf. on Plasma Physics* (Lisbon, Portugal) ECA Vol. **39E** P4.102 (4pp)
[16] Ficker O, et al. 2017 *Nucl. Fusion* **57** 076002 (10pp)
[17] Vlainic M, et al. 2015 *in Proc. 42nd EPS Conf. on Plasma Physics* (Lisbon, Portugal) ECA Vol.**39E** P4.108 (4pp)
[18] Rabinski M, et al. 2017 *J. Instrum.* **12** C10014 (7pp)
[19] Ficker O, et al. 2017 *in Proc. 44th EPS Conf. on Plasma Physics* (Belfast, Northern Ireland) ECA Vol.**41F** P5.126 (4pp)
[20] Pucella G, et al. 2015 *Nucl. Fusion* **55** 104005 (11pp)
[21] Causa F, Pucella G, B. Esposito E, Buratti P, Giovannozzi E, FTU Team, Jakubowski L, Malinowski K, Rabinski M, Sadowski M J and Zebrowski J 2015 *Proc. of Science* Vol. **ECPD2015** 066
[22] Causa F, Buratti P, B. Esposito E, Pucella G, Giovannozzi E, Jakubowski L, Malinowski K, Rabinski M, Sadowski M J, Zebrowski J and the FTU Team 2015 *Nucl. Fusion* **55** 123021(9pp)
[23] Esposito B, et al. 2017 *Plasma Phys. Control. Fusion* **59** 014044 (12pp)

Influence of gas conditions on electron temperature inside a pinch column of plasma-focus discharge

D R Zaloga[1], M J Sadowski[1,2], E Skladnik-Sadowska[1], M Paduch[2], E Zielinska[2] and K Tomaszewski[3]

[1] National Centre for Nuclear Research, A. Soltana 7, 05-400 Otwock-Swierk, Poland
[2] Institute of Plasma Physics and Laser Microfusion, Hery 23, 01-497 Warsaw, Poland
[3] ACS Ltd., Hery 23, 01-497 Warsaw, Poland

E-mail: dobromil.zaloga@ncbj.gov.pl

Abstract. The paper reports on soft x-ray emission from high-current discharges in PF-1000U facility operated at 170 kJ. The discharges at static conditions were performed with pure deuterium (D_2) and a mixture of D_2 and neon (Ne). In shots with the gas-puffing 1 cm^3 of D_2 or a (D_2+Ne) mixture was injected 2 ms before the discharge initiation. Time-integrated x-ray images from a Be-filtered pinhole camera showed that the pinch microstructure depends strongly on gas conditions. In shots with the D_2-, (D_2+Ne)- or He-puffing distinct "filaments" and "hot-spots" were observed. Time-resolved x-ray pulses were recorded with 4 filtered PIN-diodes which recorded signals from 2 regions of 3 cm in diameter (at z = 3 cm and 6 cm from the anode). From a ratio of x-ray pulses, measured behind different filters, it was estimated that at the static D_2-filling electron temperatures (T_e) were from 90 eV to 200 eV. At the D_2-filling and -puffing additional x-ray spikes were emitted from "hot-spots" with T_e twice higher. In shots at the (D_2+10%Ne)-filling T_e was 4 keV. In shots with the (D_2+Ne)-mixture puffing intense "hot-spots" were formed, and T_e reached 2.2-7.5 keV. At the same conditions "filaments" were reproducible macroscopically, but "hot-spots" were irreproducible.

1. Introduction

Studies of x-rays emitted from high-temperature plasma provide valuable information about its structure, composition and temperature [1-2]. Such studies were performed during many experiments with a modernized PF-1000U facility, but they concerned mainly time-integrated x-ray pinhole images recorded at different gas conditions [3]. More extensive studies of x-rays from PF-1000U experiments, which were carried out with two pinhole cameras, were reported in an earlier paper [4]. In those studies there were also performed time-resolved measurements with filtered PIN diodes, which observed different regions of the pinch column, but the analysis of x-ray peaks concerned only their correlations with current filaments and hot-spots visible in time-integrated x-ray images. Detailed time-integrated x-ray measurements were performed later and attention was focused on various micro-structures (current filaments and hot-spots), but the time-resolved x-ray signals were correlated with the observed current filaments only [5]. More accurate x-ray measurements in the PF-1000U facility and estimates of the electron temperature were performed recently, but analysis of time-integrated pictures was not very detailed [6-7]. Therefore, the main aim of this paper was to report on new detailed x-ray measurements carried out by means of a filtered pinhole camera and four PIN diodes, which were located behind pinholes with different filters. The first aim was to compare new time-integrated x-ray images which might supply information about differences in the pinch internal structure. The second aim was to analyze time-resolved x-ray signals (recorded behind different filters) and to estimate plasma electron temperatures.

2. Experimental setup

The described experiments were performed in a modernized PF-1000U facility which was equipped with two coaxial electrodes of 46 cm in length, oriented horizontally [8-9]. The outer electrode was composed of 12 stainless-steel tubes (each of 8 cm in diameter), distributed symmetrically upon a cylindrical surface of 40 cm in diameter. The inner electrode was made of a thick-wall copper tube of 23 cm in diameter, surrounded at the base by a ceramic insulator of 8.5 cm in length. The outlet of that electrode was closed by a thick copper plate with a central 4.5-cm-diam. hole, where was placed a nozzle of a fast gas-valve. That valve could be triggered (from an additional condenser bank) 1.5-2.0 ms before the application of a high-voltage pulse, which initiated the main PF-1000U discharge. During that initiation the inner electrode (at high potential) played a role of the anode, and the outer one was the grounded cathode. The discharges were supplied from a large condenser bank 1.33 mF, which was charged to the voltage of $U_0 = 16$ kV and stored 170-kJ energy.

Plasma discharges were investigated with x-ray framing cameras and x-ray pinhole cameras recording time-integrated x-ray images. For time-resolved studies of x-rays the use was made of PIN-type semiconductor detectors, which were placed behind filtered pinholes [10]. There was also used a four-frame camera with a microchannel plate (MCP) which recorded UHV and XUV images. An arrangement of the applied diagnostic equipment is shown in figure 1a.

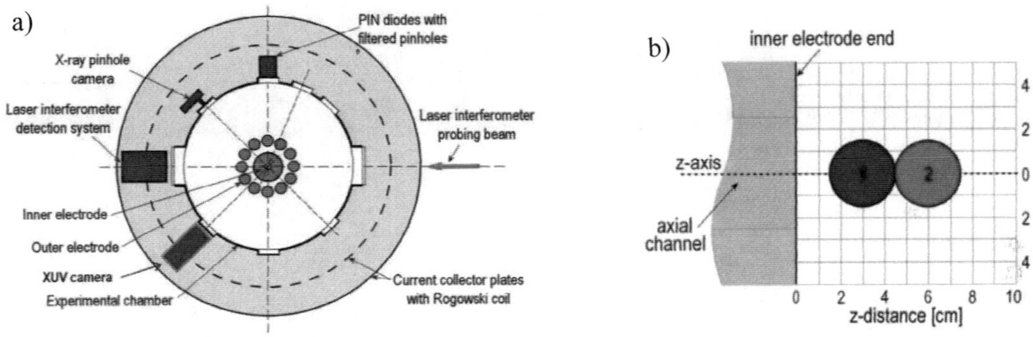

Fig. 1. (a) Cross-section of the PF-1000U chamber, which shows positions of the diagnostic equipment, (b) observation fields (each for two PIN diodes) for time-resolved x-ray measurements.

During all investigated discharges the total fusion-neutrons yield (Y_n) was measured by means of several silver-activation counters located around the PF-1000U chamber, and time-resolved measurements of hard x-rays and fusion-neutrons were performed with several scintillation probes placed at different distances from the electrodes outlets.

In the reported study time-integrated x-ray images of plasma were recorded with x-ray films placed inside a pinhole camera, which was oriented at 75° to the z-axis and looked at the inner electrode outlet and pinch column of about 10 cm in length. It was equipped with an input 500-μm diaphragm covered by a 10-μm Be-filter, which enabled to record x-rays of energy > 1 keV.

To record time-resolved x-ray signals from a pinch column, the use was made of four PIN-diodes of the Hamamatsu-S9055 type, which were placed behind a diaphragm with four 100-μm pinholes covered by different absorption filters. The whole set of the PIN diodes, produced by the ACS Ltd, Poland, was situated in a small vacuum chamber pumped out by a turbo-molecular pump [10]. The diodes windows were removed to detect photons < 1 keV. These PIN-diodes were shielded by 7-μm or 10-μm Be-filters. Each pair of the diodes observed another 30-mm-diam. region of the pinch column, as shown in figure 1b.

To estimate the plasma electron temperature (T_e) it was taken into account that recording the x-ray emission at a wavelength λ (and energy $E = hc/\lambda$) behind a filter (j) the total intensity of the x-radiation depends on T_e and is given by the known formula [11]

$$I_\lambda \propto \lambda^{-2} \left({}^{k}/_{T_e} \right)^{-1/2} \exp\left[{}^{-E}/_{kT_e} - \sum \mu_j(E) \times d_j \right]$$

where the subscript j describes a specific material, $\mu_j(E)$ is the absorption coefficient of the filter material, and d_j is its thickness. Since the used absorption filters were made of the same material (Be) the summation was not needed. High-temperature plasma without considerable impurities was considered, and under assumption that T_e are high enough (to neglect recombination and linear x-ray emission) the Bremsstrahlung emission (characterized by the Maxwellian distribution) was taken into account only. Those assumptions were consistent with results of the OES measurements [12], which showed a small amount of impurities lines during the maximum compression (t = 0). They agreed also with results of computations of a short thermalization time for a dense plasma column. Hence, using two identical detectors (PIN-diodes denoted by 1 and 2) which observed the same plasma region, but were shielded by Be-filters of a different thickness (d_1 and d_2), it was possible to estimate T_e from the ratio of x-ray emission intensities (I_1/I_2) measured behind these filters [11].

Since the reported measurements were performed with the PIN-diodes, which were located behind the identical pinholes covered by the 7-µm and 10-µm Be-filters, it was possible to estimate T_e values from a diagram showing a dependence of the I_1/I_2 ratio on a T_e value, as presented in figure 2.

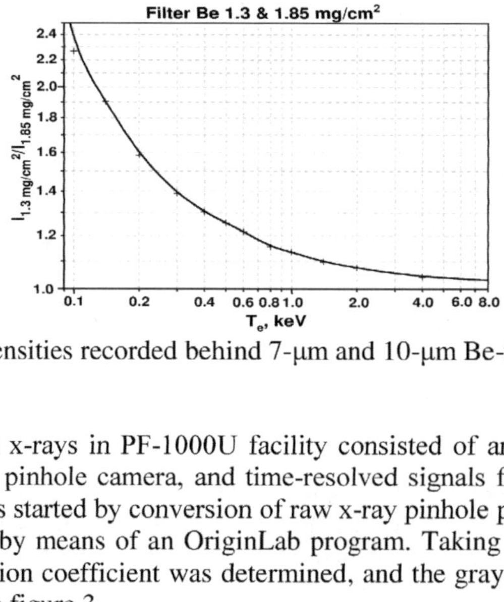

Fig. 2. Ratio of the x-ray intensities recorded behind 7-µm and 10-µm Be-filters, as a function of T_e.

3. Experimental results

The reported studies of soft x-rays in PF-1000U facility consisted of an analysis of time-integrated images from the Be-filtered pinhole camera, and time-resolved signals from the filtered PIN diodes. The preliminary analysis was started by conversion of raw x-ray pinhole pictures into colour-enhanced equidensity-contour images by means of an OriginLab program. Taking into account the experiment configuration the magnification coefficient was determined, and the gray-scale was converted into 17 colour-levels, as presented in figure 3.

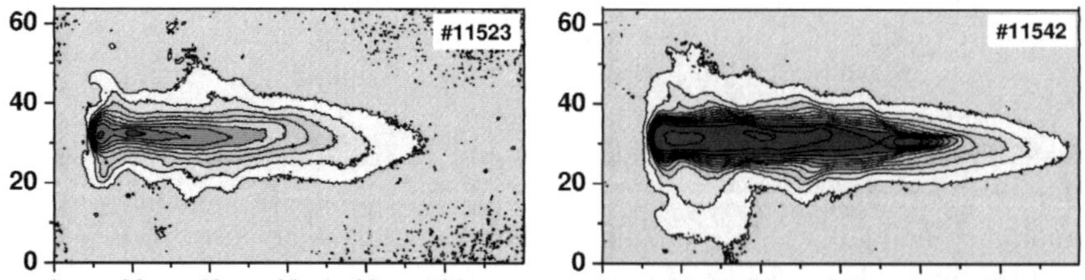

Fig. 3. Colour-enhanced equidensity images of the X-rays emission from discharges: #11523 (p_o = 1.2 hPa D_2, U_0 = 16 kV, Y_n = 1.1 x 10^{10}), and #11542 (p_o = 1.2 hPa D_2 + D_2-puffing, Y_n = 9.6 x 10^{10}).

From the images presented above one could easily deduce that the use of the additional D_2-puffing induced a considerable increase in the soft x-rays emission. The length of the plasma jet was about 10 cm, but there were observed some microstructures, which probably corresponded to hot-spots.
For the discharges imaged in figure 3 there were also recorded time-resolved signals from the PIN diodes described in Section 2. On the basis of those signals and the diagram presented in figure 2 it was possible to estimate values of the local electron temperature (T_e), as shown in figure 4.

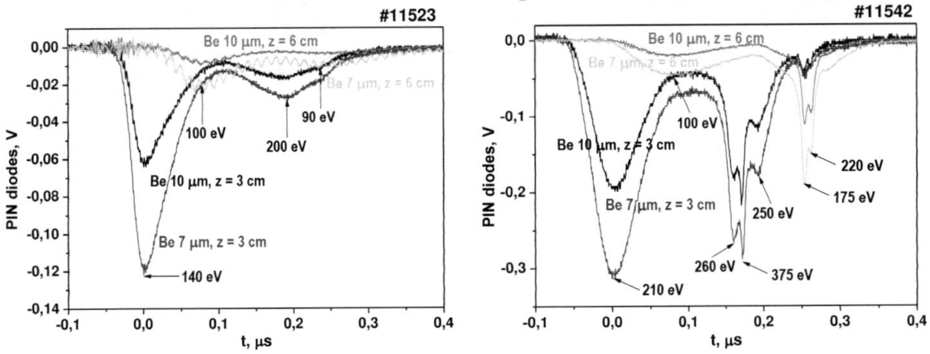

Fig. 4. Soft x-ray signals from two observation regions, as recorded for discharge #11523 (performed at p_o = 1.2 hPa D_2 and U_o = 16 kV) and for discharge #11542 (performed with the additional D_2-puffing), and values of the local T_e.

The results presented above showed that in the discharge #11523 the local T_e values in the 1st observation region (with the centre at z = 3 cm) changed from 140 eV to 200 eV, while those in the 2nd region (with the centre at z = 6 cm) decreased from 100 eV to 90 eV. In the discharge #11542 the local T_e values were higher, in the 1st region they changed from 210 eV 375 eV, and in 2nd region they rose from 100 eV to 220 eV. The narrow spikes observed during the second x-ray pulse, particularly for shot #11542, might be explained as signals from hot-spots with noticeable higher T_e. All the T_e values (estimated with an experimental error of ±30%) seemed to be reasonable, taking into account the results of earlier studies of PF-type discharges [6].
Analogous measurements of the soft x-ray emission were also carried out for discharges performed with a (D_2+Ne)-filling, as shown in figure 5.

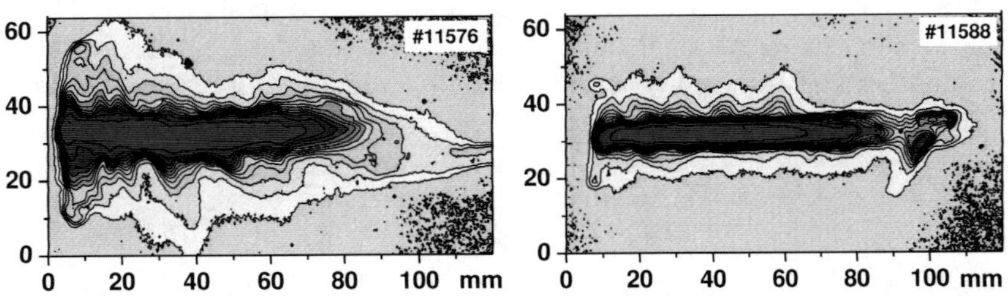

Fig. 5. Colour-enhanced equidensity images of X-rays emission from two discharges at the initial filling: p_o = 1.2 hPa (90%D_2 + 10%Ne) and U_0 = 16 kV: #11576 (with no additional gas-puffing, Y_n = 3.9 x 10^{10}), and #11588 (with additional puffing of a (75%D_2 + 25%Ne) mixture, Y_n = 3.1 x 10^{11})

The use of the (90%D_2 + 10%Ne) mixture caused also an evident increase in the soft x-ray emission. The application of the additional puffing of the (75%D_2 + 25%Ne) mixture did not induced any considerable change of the plasma column length, but it caused a decrease in the pinch diameter, and the formation of more distinct hot-spots. It should be noted that such effects were also observed for other discharges performed under the similar gas-conditions, but the image of shot #11588 presented the pinch column of the smallest diameter.

For the discharges imaged in figure 5 time-resolved x-ray signals from the PIN diodes were also recorded. On the basis of those signals and the diagram shown in figure 2 it was possible to estimate values of the local T_e, as shown in figure 6.

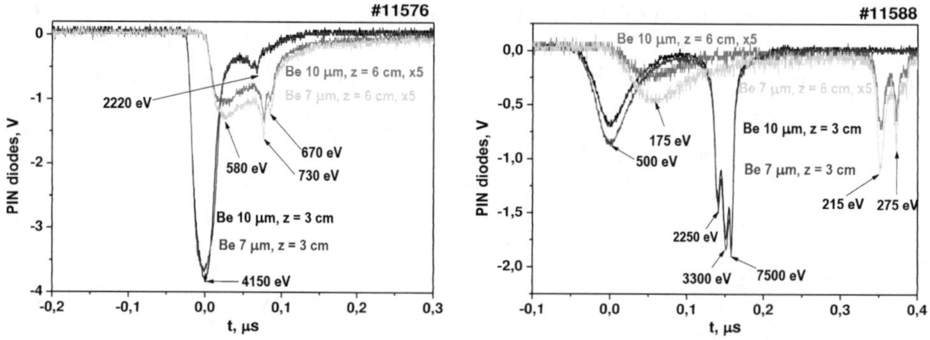

Fig. 6. Soft x-ray signals from two observation regions, as recorded for discharge #11576 (performed p_o = 1.2 hPa (90%D_2 + 10%Ne), U_o = 16 kV), and for discharge #11588 (performed with additional injection of the (75%D_2 + 25%Ne) mixture), and corresponding values of the local T_e.

The results presented above showed that for discharges with the (90%D_2 + 10%Ne) filling the structure of the recorded x-ray signals was more complicated, i.e., the number of the observed spikes was increased, and the estimated T_e values were considerably higher. In the discharge #11576 the local T_e values in the 1st observation region decreased from 4.15 keV to 2.2 keV, while those in the 2nd region changed from 580 eV to 730 eV. In the discharge #11588 (with the additional gas-puffing) T_e values in the 1st observation region changed from 0.5 keV even to 7.5 keV, while those in the 2nd region changed from 175 eV to 275 eV. It should, however, be noted that for discharges with the use of a Ne-admixture the T_e estimation errors were much larger (reaching even ±80%) because of two reasons. The first one was fact that the thickness of two filters was too small to estimate T_e in the optimal range on the curve presented in figure 2. The second reason was the appearance of a very intense Ne-line (of energy equal to about 1 keV) which induced a strong increase in the soft x-ray emission, and larger inaccuracy of T_e estimations.
During the recent experimental session there were also performed discharges with the pure D_2-filling and the additional He-puffing [13]. For the discharge #11830 the use was made of the XUV camera, which recorded frame-pictures, each with the exposition time equal to about 2 ns. The conversion of those white-black pictures into colour-enhanced images (as it was done earlier for the x-ray pinhole pictures) made it possible to observe a very complicated filamentary structure and numerous hot-spots, as shown in figure 7.

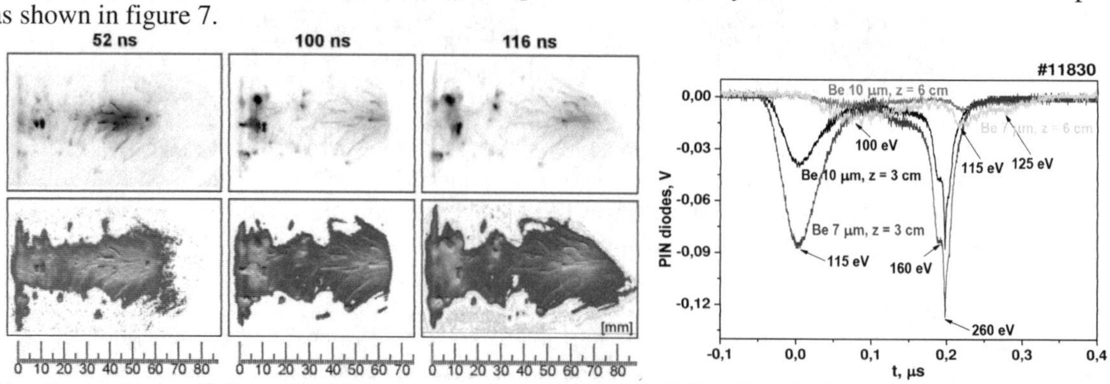

Fig. 7. Analysis of three XUV frame-pictures, as recorded for the discharge #11830, which was performed at p_o = 1.1 hPa D_2 + the additional He-puffing, U_0 = 16 kV, and enabled the dimensions of the plasma filaments and hot-spots to be evaluated, and the time-resolved soft x-ray signals from the same discharge, which show corresponding values of the local T_e.

IOP Conf. Series: Journal of Physics: Conf. Series **959** (2018) 012003 doi:10.1088/1742-6596/959/1/012003

The results obtained for shot #11830 showed that for discharges with the pure D_2-filling and He-puffing there were also observed some spikes which corresponded to hot-spots visible in the XUV pictures. The estimated T_e values were lower than those for discharges with the D_2-puffing (figure 4). The T_e values in the 1st observation region changed from 115 eV to 160 eV, but there was also a spike characterized by 260 eV. The T_e values in the 2nd region were equal to 100-125 eV only.

4. Summary and conclusions

The most important results of the reported studies can be summarised as follows: Plasmas generated in the PF-1000U facility were studied under different gas conditions. Time-integrated x-ray pinhole pictures, time-resolved x-ray signals from PIN diodes, and XUV images showed that the pure D_2-discharges (with and without any gas puffing), as well as those performed with the (D_2 + Ne) filling and (D_2, Ne or He)-puffing, produced dense plasma emitting intense x-rays. The recorded time-resolved x-ray signals made it possible to determine a ratio of x-rays intensities (I_1/I_2) for two observed plasma-regions. From that ratio it was estimated that: For the pure D_2-shots the T_e values were 140-200 eV in the 1st region and about 100 eV in the 2nd region, and for the D_2-shots with D_2-puffing those values amounted to 210-375 eV and 100-220 eV, respectively. For the D_2-shots with He-puffing those values were lower and reached 115-160 eV and 100-125 eV only. When the (90%D_2 + 10%Ne)-filling was applied the T_e values in the considered regions amounted to 4.15-2.2 keV and 580-730 eV, respectively, and when the (75%D_2 + 25%Ne)-puffing was added those values reached even 7.5 keV. Estimation errors for lower T_e values were about ±30%, but those for higher T_e might amount even ±80%. The reasons of this difference were too small thicknesses of the Be-filters and inaccuracies in the determination of x-ray peaks profiles. To determine I_1/I_2 values the x-ray signals should be integrated, but for narrow x-ray pulses one might use their amplitudes ratio.

In conclusions - it should be noted that: 1. The estimated high T_e values (observed particularly for discharges with the additional gas-puffing) corresponded evidently to filaments and hot-spots which could be quasi-stable for a relatively long period, as was shown in earlier PF studies [14-15]; 2. At the pure D_2-filling the use of the He-puffing did not rise T_e because the x-ray emission intensity depended not only on amount of electrons (linked to Z value), but also on the plasma compression and density, which determined the electrons deceleration and the Bremssstrahlung emission; 3. The application of a heavier admixture (e.g. Ne) in the filling and puffing induce evidently the formation of more plasma microstructures; 4. The obtained high T_e values (up to 7.5 keV) are consistent with those observed in other PF experiments, where T_e ranging to 5-10 keV were found [6, 16].

References

[1] *Plasma Diagnostic Techniques*, Edit. R.H. Huddlestone, S.L. Leonard (Academic Press, 1965).
[2] M. J. Sadowski, and M. Scholz, Plasma Sources Sci. Technol. **17** (2008) 024001.
[3] P. Kubes, M. Paduch, et al., Phys. Plasmas. **21** (2014) 082706.
[4] W. Surala, M. J. Sadowski, et al., Nukleonika **60** (2015) 303-308.
[5] M. J. Sadowski, M. Paduch, et al., Plasma Sources Sci. Technol.**24** (2015) 055003.
[6] E. Skladnik-Sadowska, D. Zaloga, et al., Plasma Phys. Control. Fusion **58** (2016) 095003.
[7] E. Skladnik-Sadowska, M. J. Sadowski, et al., PAST Ser. Plasma Phys. No **6** (2016) 112-116.
[8] M. Scholz, L. Karpinski, et al., Nukleonika **57** (2012) 183-188.
[9] D. Zaloga, E. Skladnik-Sadowska, et al., Nukleonika **60** (2015) 309-314.
[10] *SXRDS Soft X-ray Detection Set; Abridged Description* (ACS Ltd. Warsaw, 2015).
[11] R. C. Elton, NRL Report 6738 (NRL, Washington, DC, USA, 1969).
[12] E. Skladnik-Sadowska, R. Kwiatkowski, et al., PAST Ser. Plasma Phys. No **1** (2013) 279-283.
[13] D. R. Zaloga, *Research on emission of visible and X-ray radiation, and estimations of electron temperature in discharges of plasma-focus type*, PhD Thesis (NCBJ, Swierk 2017) – in Polish.
[14] M. Sadowski, H. Herold, et al., Phys. Letters **105A** (1984) 117-123.
[15] P. Silva, and M. Favre, J. Phys. D: Appl. Phys. **35** (2002) 2543.
[16] W. L. Harries, Final Technical Report NASA-CR-1132500 (Old Dominion Univ., Norfolk 1974).

IOP Conf. Series: Journal of Physics: Conf. Series **959** (2018) 012004 doi:10.1088/1742-6596/959/1/012004

Materials surface damage and modification under high power plasma exposures

I Garkusha[1,2], V Makhlaj[1,2], O Byrka[1], V Taran[1], V Voitsenya[1], S Malykhin[3], S Herashchenko[1], S Surovitskiy[3], K Nowakowska-Langier[4], M J Sadowski[4], E Skladnik-Sadowska[4], D. Terentyev[5]

[1]National Science Center "Kharkiv Institute of Physics and Technology", Kharkiv, Ukraine

[2]V. N. Karazin Kharkiv National University, 61022 Kharkiv, Ukraine

[3]National Technical University "Kharkiv Polytechnical Institute", Kharkiv, Ukraine

[4]National Centre for Nuclear Research (NCBJ), 05-400 Otwock, Poland

[5]Nuclear Materials Science Institute, SCK•CEN, Mol, Belgium

garkusha@ipp.kharkov.ua

Abstract. Influence of powerful plasma impacts on several materials used for the construction of energy systems, i.e. different grades of steels as well as tungsten coatings, has been discussed. Irradiations of these materials with hydrogen and helium plasma streams have been performed in several high-current-pulse and quasi-stationary plasma accelerators which provided the variation of a power load upon the exposed surface as well as changes of the particle flux in wide ranges: the energy flux density in the range of 1-25 MJ/m^2, particle flux - up to 10^{26}-10^{29} ion/m^2s, the plasma stream velocity - up to about 500 km/s, and the pulse duration in the range of 1-250 μs.

A response of the investigated materials to extreme plasma loads, which are relevant to transient events in fusion reactors, is briefly discussed. It is demonstrated that a broad combination of mechanisms of powerful plasma interactions with various materials includes not only a surface damage caused by different erosion mechanisms, but under certain conditions it may also result in a significant improvement of material properties in the near-surface surface layer of several tens-μm in thickness. Some improvement of the structure and substructure of such a layer may be caused by the high-speed quenching, the shock wave formation and material alloying with plasma- and coating-species. The creation of unique surface structures and a considerable improvement of physical and mechanical properties of different materials can be achieved by the pulsed plasma alloying, i.e. pre-deposited coating modifications and mixing caused by the impacting plasma streams.

1. Introduction

Some low-activation steels are considered as constructional materials for nuclear and fusion devices due to their high swelling resistance and low irradiation creep-rates (i.e. slow changes in dimensions of materials exposed to prolonged stresses, caused by X-rays, γ-rays and neutron irradiation, etc.). In particular, such steels might be a possible option at the choice of plasma-facing materials for the first wall surfaces in DEMO [1-3].

The main drawback of such steels at their application as the plasma facing materials is their high sputtering rate under influence of energetic ions (particularly hydrogen isotopes). It is directly related with the generation of impurities as well as with a lifetime of the plasma facing components [3-4]. One of the prospective ways towards an improvement of the steel properties is the alloying of their surface layers with some refractory elements [3]. Plasma ions can of course be considered as a source of the alloying elements which should be introduced into a modified layer structure [5]. Another possibility of the alloying during the pulsed plasma processing is the mixing of previously deposited thin ($h_{coat} < h_{melt}$) coatings of a different pre-determined composition with the chosen substrate in the course of its melting driven by powerful plasma impacts [6-7].

The alloying of a surface layer in result of the coating-substrate mixing allows achieve a desirable chemical composition of the processed surface layers [5-9]. Nevertheless, the modification processes in plasma facing materials, which can be induced by repetitive plasma impacts and synergetic effects (caused by different factors) are still not understood satisfactorily to make reasonable predictions for steels behavior during the fusion-reactor operation. All the mentioned issues require comprehensive studies of prospective steel materials with different powerful plasma devices (simulators) and coordinated activities, which should address relevant issues - associated with the impact of intense heat and particle loads on the investigated plasma facing materials.

Analyses and comparisons of experimental results of research on materials behaviour under their irradiation with steady-state and high-power pulsed plasma-streams, which can be generated in different plasma devices with various specific heat loads and different particle fluxes, are necessary. They can give the unique possibility to understand the important common features of surface damages at extreme conditions, to investigate the peculiarities of the surface modification of low-activation steels and possible erosion mechanisms in dependence on the plasma parameters and the impact load.

This paper presents experimental studies of the alloying and modification of the chosen commercial grades of steels, and particularly the (Cr18Ni10Ti) austenitic steel and ferritic/martensitic Eurofer alloy (Cr - 9.7%, Mn - 0.7%, Fe - 89.6%) with some tungsten admixture introduced by the plasma-induced mixing. There are also discussed studies of the sputtering of the surfaces covered by tungsten and exposed to powerful plasma streams.

2. Samples, experimental devices and diagnostics

In the reported experiments there were applied planar samples (of dimensions of $20\times20\times1$ mm^3), made of the well-known austenitic steel of the Russian trademark (Cr18Ni10Ti), which has the following composition (18 wt% of Cr, 10 wt% of Ni, 1 wt% of Ti, and the rest - Fe). Such a steel is an analog of the stainless steel marked as AISI 321. The samples made of the Eurofer alloy (Cr-9.7%, Mn-0.7%, Fe-89.6%) were prepared and delivered by the SCK•CEN laboratory (Belgium). All the samples had the size of $10\times10\times0.5$ mm^3. These samples were covered by tungsten coatings of about 3 μm in thickness, which were deposited with a PVD technique [7] within a Bulat-type facility. Earlier such coatings were applied for estimations of their performance as plasma-facing surfaces (in a comparison with monolithic tungsten targets). They enabled also an analysis of the adhesion properties of the PVD coatings, and an investigation of the plasma-facing components prospective for the reactor first wall construction [10].

Surface modifications by powerful pulsed plasma streams were carried out with the use of a QSPA Kh-50 quasi-stationary plasma accelerator and a MPC magneto-plasma compressor [10-12]. The main parameters of the QSPA hydrogen plasma streams were as follows: ion impact energy was about (0.4…0.6) keV, the maximum plasma pressure reached up to 0.32 MPa, and the plasma stream diameter was about 18 cm. The surface energy loads, as measured with a calorimeter, amounted to 0.6 MJ/m^2 (i.e. near the tungsten melting threshold). The load pulse shape was approximately triangular, and the pulse duration was 0.25 ms [10-12]. The MPC facility generated compressed plasma streams of the density amounting to about 10^{18} cm^{-3}, and a plasma energy density was varied in the range of 0.05–0.5 MJ/m^2. The amplitude of the discharge current in the accelerating channel amounted to about 500 kA, and the discharge half-period was equal to about 10 μs. The experiments were carried out using pure helium at the initial pressure equal to 266.64 Pa. The surface energy loads, as measured with calorimeters, amounted to 0.4 MJ/m^2. During the reported experiments, the target surface before each plasma pulse was maintained at the room temperature.

The sputtering procedures were performed within the DSM-2 stand described in an earlier paper [13]. An ion source was the system based on a microwave discharge in argon within a mirror magnetic trap, which was operated at the electron cyclotron resonance (ECR) frequency of 2.37 GHz. The plasma density n_e in this system was about 10^{10} cm^{-3}, and the electron temperature T_e reached about 5 eV. According to measurements, performed with an Omegatron-type gas analyzer, the main impurity was water, but the total impurity amount was negligibly in a comparison with the argon content. The plasma ions were accelerated by a negative potential (-600 V) applied to the water-cooled holder. The density of the ion current upon the mirror surface was equal to 1.6 mA/cm^2.

IOP Conf. Series: Journal of Physics: Conf. Series **959** (2018) 012004 doi:10.1088/1742-6596/959/1/012004

The energy density in free plasma and the surface heat loads were measured by means of the local calorimeters. Observations of plasma interactions with the exposed surfaces were performed with a high-speed 10-bit CMOS pco.1200s digital camera of the PCO AG type (in the spectral range from 290 nm to 1100 nm, with an exposure time ranging from 1 µs to 1 s) [10-12]. The surface analysis of the exposed samples was carried out with a MMR-4 optical microscope, equipped with a CCD camera. There were also performed measurements of weight losses, as well as precise measurements of the surface roughness with a Hommelwerke T500 tester. To study a micro-structural evolution and contents of elements and phases in the exposed targets, the x-ray diffraction technique (XRD) was applied [14-17].

3. Experimental results

3.1. Initial state of sample surfaces

The values of the initial micro-hardness and roughness of the Cr18Ni10Ti samples amounted to 350 kg/mm^2 and $R_a < 0.1\ \mu m$ ($R_{max} \approx 0.1$ µm), respectively. Upon the virgin surfaces only some lines corresponding to the γ-Fe phase were observed [17]. It should be mentioned that a face-centred cubic lattice was attributed to that phase.

For the Eurofer samples the initial micro-hardness and roughness were 190 kg/mm^2 and $R_a < 0.1\ \mu m$ ($R_{max} \approx 0.1$ µm), respectively. Surface profiles of these samples are shown in Figure 1.

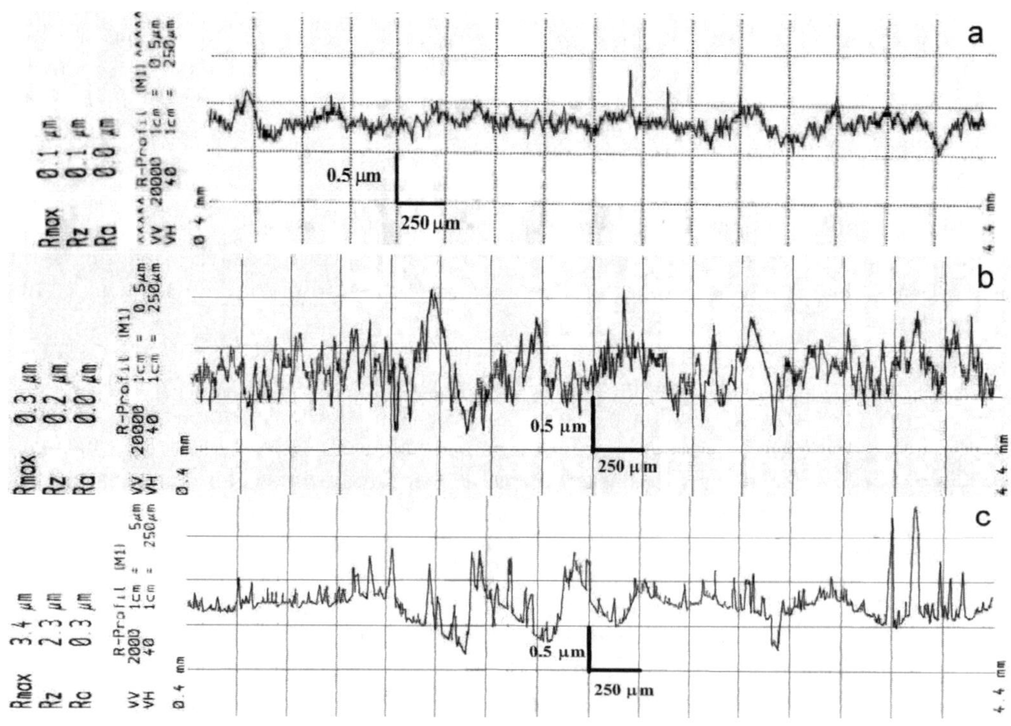

Figure 1. Surface profiles of the Eurofer samples: (a) for a virgin sample, (b) for that exposed to 5 QSPA-plasma pulses, and (c) for that covered by a W-layer and exposed to QSPA plasma. In those cases the heat load was Q = 0.6 MJ/m^2.

The composition the Eurofer samples in the initial (virgin) state and after their processing is shown in Table 1.

During structural studies of the samples surfaces there were observed lines corresponding to the α-Fe phase only. It should be noted that to that phase there was attributed the body-centred cubic crystal structure. The results the diffraction measurements are presented in Figure 2.

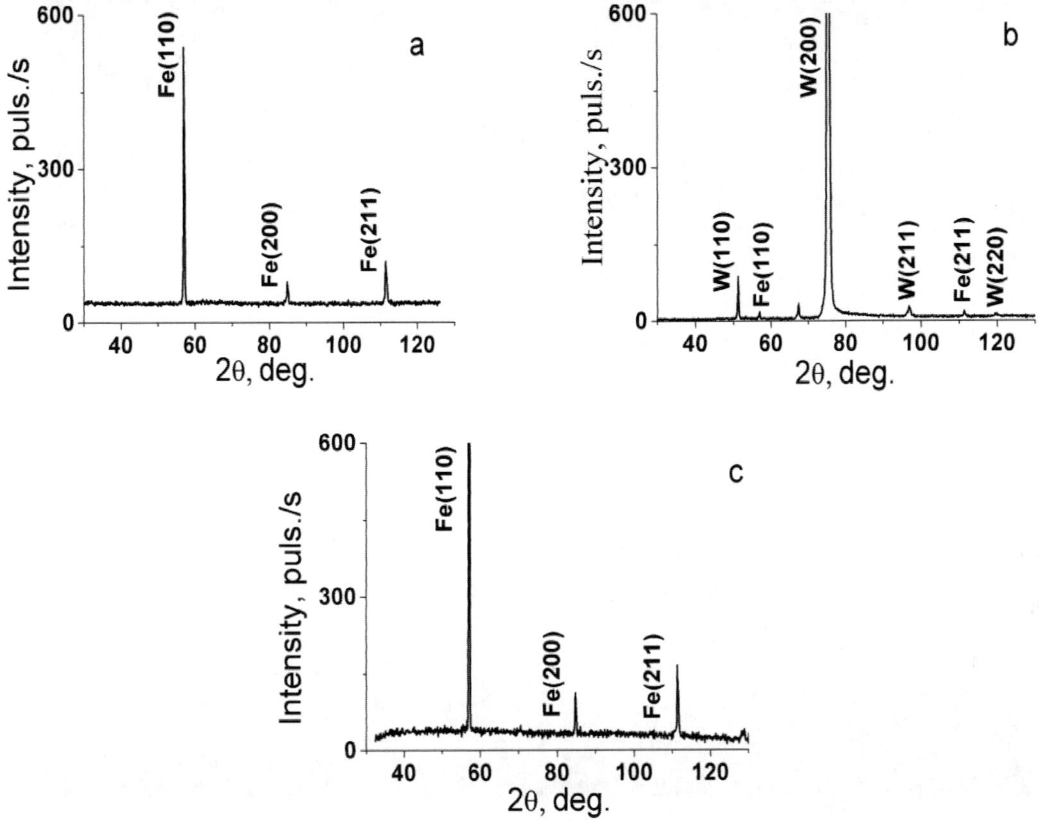

Figure 2. Diffraction patterns (Fe-K α1 radiation) of the Eurofer samples: (a) measured in the initial state, (b) measured for the sample coated with a W-layer and exposed to QSPA-plasma pulses, and (c) measured for the sample coated with a W layer, modified by QSPA-plasma pulses and irradiated by the Ar-ion$^+$beam.

Table 1. Element content of the investigated Eurofer samples

Sample	**Cr** (wt %)	**Mn** (wt %)	**Fe** (wt %)	**W** (wt %)
Initial	9.7	0.4	89.9	-----
Coated with W and exposed to QSPA-plasma	1.3	----	12.95	85.8
Coated with W, exposed to QSPA-plasma and sputtered by the Ar-ion beam	8.9	0.36	89.64	1.1
Coated with W and exposed to MPC-plasma	9.7	0.4	88.8	1.1
Coated with W and sputtered by the Ar-ion beam	9.6	0.4	88.9	1.1

3.2. Surface modification by powerful plasma

Irradiations of the Cr18Ni10Ti-steel and Eurofer-alloy samples by plasma streams within the QSPA and MPC facilities led to the surface melting and resulted in the development of re-solidified surface layers, which are presented in Figures 3 and 4.

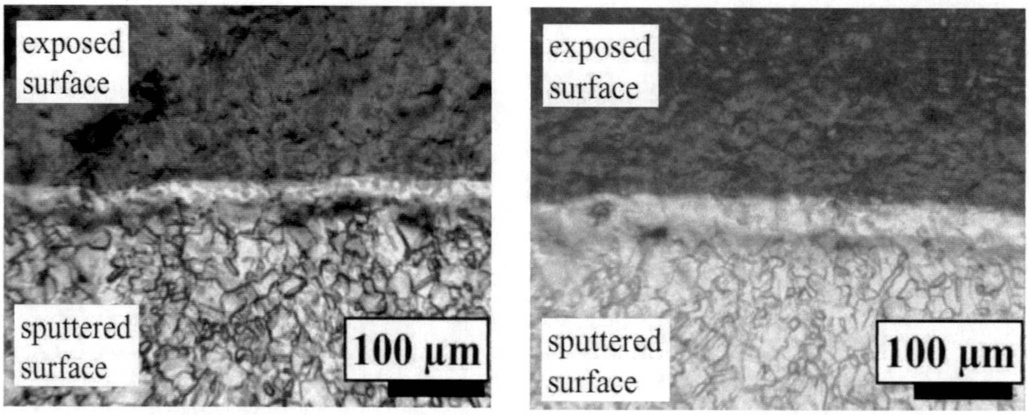

Figure 3. Optical microscope images of the Cr18Ni10Ti-steel surface exposed to 5 plasma pulses within the MPC (left) and QSPA (right) facilities. The lower parts of both images show the same modified surfaces after their sputtering by the Ar^+ beam. The black marker-stripe length is 100 μm.

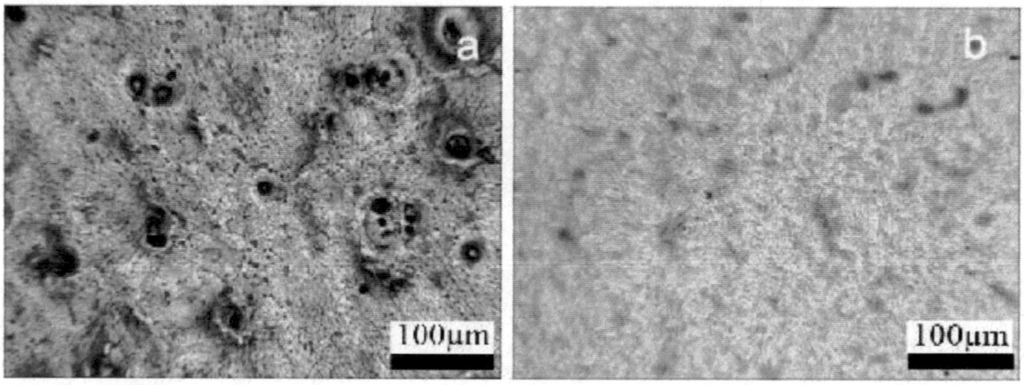

Figure 4. Optical microscope images of the Eurofer-alloy surface exposed to 5 plasma pulses within the MPC (a) and QSPA (b) facilities. The black marker-strip shows the scale.

The surface morphology was changed mainly due to the formation of the plasma-melted layers. The roughness of the exposed surfaces was increased up to $R_{max} \approx 0.3$ μm (see Figure 1b). The α-Fe phase was identified together with lines corresponding to the γ-Fe phase upon the treated surfaces of the Cr18Ni10Ti samples. In contrary the content of phases upon the Eurofer surface was not changed.

It should be noted that the micro-hardness of the modified surface of the Cr18Ni10Ti sample as well as that of the Eurofer sample was increased up to 400 kg/mm^2 and 260 kg/mm^2, respectively, due to the stresses and quenching induced by the plasma treatment [18].

3.3. Modification of tungsten coatings deposited on steel substrates

The process of the steel-surface alloying consisted of two stages. During the first stage the tungsten coating was deposited upon the sample surface by means of the PVD method. After the deposition of the tungsten coatings on the treated Cr18Ni10Ti and Eurofer surfaces their roughness was slightly increased. During the second stage the W-coated samples were exposed to 5 plasma pulses in the QSPA or MPC device.

After the irradiation of the Cr18Ni10Ti sample by plasma within the MPC or QSPA device the surface morphology was changed due to the formation of macro- and micro-cracks as well as some pores, as shown in Figures 5 and 6.

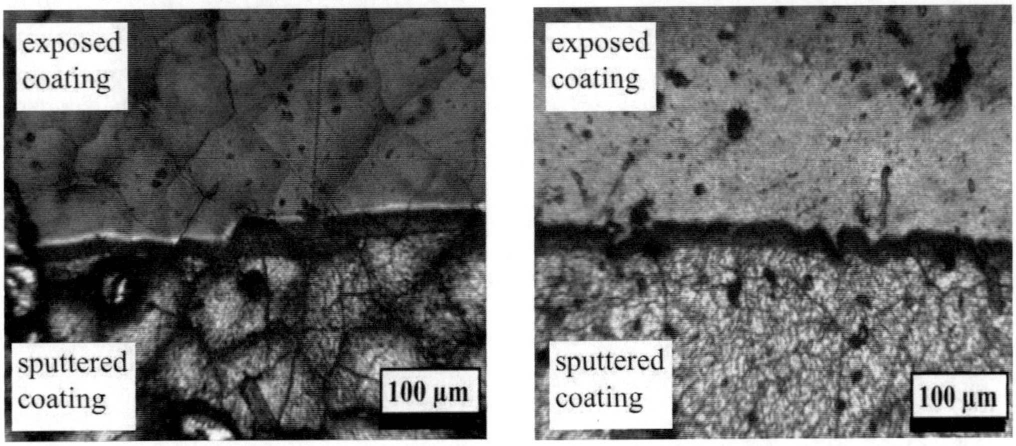

Figure 5. Optical microscope images of the Cr18Ni10Ti-steel surface coated by a W-layer and exposed to 5 plasma pulses in the MPC (left) and QSPA (right) facilities. The lower parts of both images show the same modified surfaces after their sputtering by the Ar^+ beam. The length of the black marker-stripe is 100 μm.

The roughness of the exposed surfaces increased up to $R_a \approx 0.3$ μm, $R_{max} \approx 3.4$ μm (see Figure 1c), and the intensity of tungsten lines was considerably higher than that of the substrate lines (see Figure 2b). For the Cr18Ni10Ti samples the delamination of the deposited coatings was not observed (see Figure 4). An increase in the tungsten concentration (up to several wt %) was observed upon the modified surfaces.

The delamination of the coatings upon the Eurofer surfaces was not observed during their irradiation by plasma streams within the QSPA facility. The concentration of tungsten in the surface layer of 4 μm in depth achieved 85.8 wt% (see Table 1). In contrary, the Eurofer surfaces coated by W-layers and irradiated by plasma in the MPC device showed some delamination of these coatings, as shown in Figure 6.

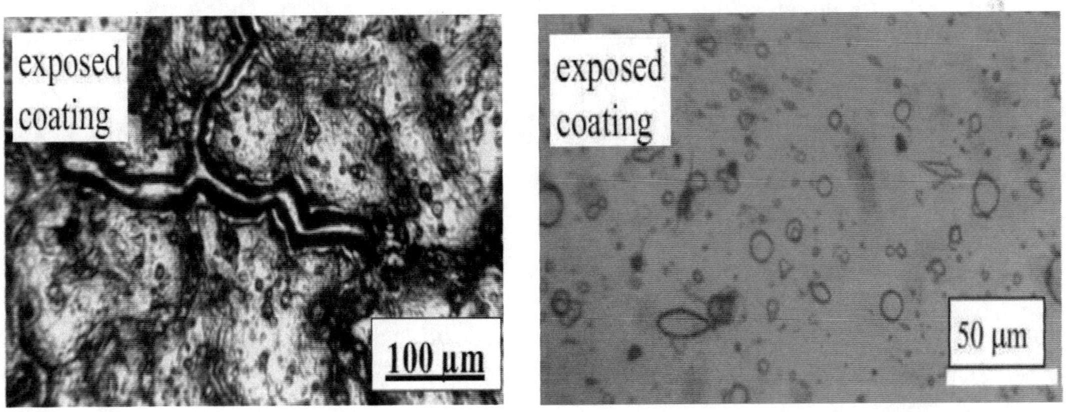

Figure 6. Optical microscope images of the Eurofer-alloy surface coated by a W-layer and exposed to 5 plasma pulses within the MPC (left) and QSPA (right) facilities.

In the considered case the concentration of tungsten in the modified layer was below 1.1 wt % (see Table 1). It should be mentioned that changes of the micro-hardness of the modified coatings upon the Cr18Ni10Ti and Eurofer samples were not very different from those observed during the modification of pure steel surfaces.

3.4. Sputtering tests of steels

The sputtering tests of the Cr18Ni10Ti and Eurofer samples were carried out for the virgin (initial) surfaces, as well as for those coated by a W-layer and those modified by plasma streams of different pulse length. The roughness of the initial surfaces was noticeably increased due to the sputtering of boundaries of the grains, as shown in Figures 3 and 7.

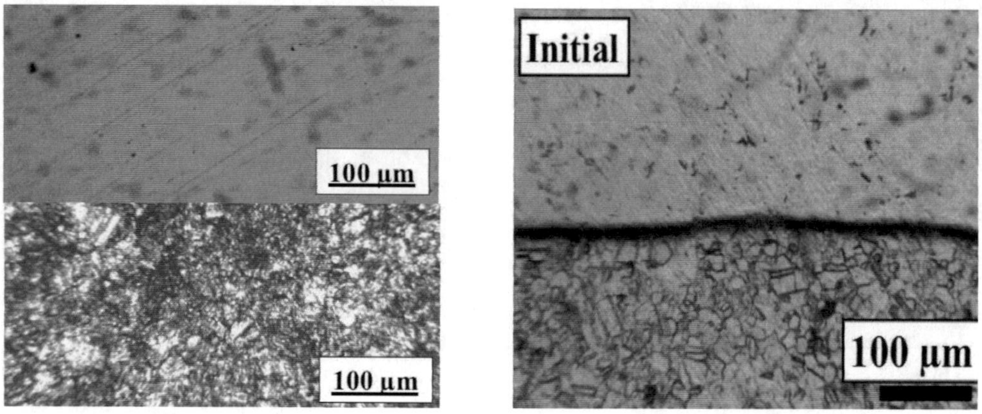

Figure 7. Optical microscope images of the Eurofer (left) and Cr18Ni10Ti steel (right) surfaces, as obtained for the initial virgin state (top parts of the pictures) and for the same surfaces after their sputtering by the Ar^+ beam (the lower parts of the pictures).

The roughness of the surfaces of the both grades of steel, which were coated by a W-layer and modified by the pulsed plasma streams, was decreased due to the sputtering of non-uniform melted and re-solidified layers and delamination of the coatings, as shown in Figures 5 and 8.

It should be noted that practically identical mass losses were measured for the steel surfaces coated by a W-layer, and for those covered with a W-layer and subsequently modified by plasma streams.

Figure 8. Optical microscope images of the Eurofer surfaces coated by a W-layer, exposed to 5 plasma pulses in the MPC (left) and QSPA (right) facilities, and finally sputtered by the Ar^+ beam.

4. Discussion and conclusions

The most important results of these studies can be summarized as follows: Experimental research on surface modifications of different stainless-steel (SS) grades was performed. These SS-grades are a possible options as plasma-facing materials of the DEMO-reactor first wall. The samples of the Cr18Ni10Ti austenite- and Eurofer (Cr-9.7%, Mn-%0.4, Fe-89.6%) ferritic/martensitic-steels covered by tungsten coatings were treated within the QSPA Kh-50 quasi-stationary plasma accelerator and the MPC magneto-plasma compressor. The heat load upon the sample surface was near the melting threshold (i.e. about 0.6 MJ/m^2, with pulse duration τ = 0.25 ms in the QSPA Kh-50, and about 0.4

MJ/m^2, with pulse duration $\tau = 20$ μs in the MPC). The preliminary tungsten coatings were deposited by a PVD method. The sputtering tests of the modified surfaces were also performed.

Possibility of the alloying of SS-surfaces with tungsten coatings was demonstrated. An increase in the tungsten concentration was observed. The tungsten phase was identified together with some lines of the Fe phase upon the treated surfaces. The presence of the α-Fe phase created good conditions for the tungsten penetration into the affected layer. The tungsten concentration achieved several wt% in the surface layer of thickness up to 4 μm. The maximum tungsten content of about 85 wt% was observed in the surface layers of the Eurofer samples modified by plasma streams in the QSPA. Nevertheless, symmetrical tensile stresses up to 270 MPa were recorded in the near surface layer. As a result of the stresses development, some delamination of the coatings during the pulsed plasma irradiation was observed. The surface morphology was changed mostly by the melting and re-solidification of a surface layer. Macro- and micro-cracks appeared also on the modified surfaces.

The sputtering yield for the SS-samples modified by plasma streams was not very different from that observed for the virgin samples. An explanation of such a surface behaviour might be the accumulation of elastic energy in the stressed surface-layer, and not very good adhesion of the modified tungsten coatings to the steel substrates. As a result of those effects some delamination of the coatings could develop. A possible way to improve the coatings resistance might be the application of several cycles of the plasma treatment [6]. During the first stage of such a cycle the tungsten coating of 1-2 μm in thickness should be deposited on the sample surface by the PVD method. During the second stage the coated samples should be processed with pulsed plasma streams. The reduction of the coating thickness together with an increase in the number of plasma treatment cycles might create conditions for the better penetration of an alloying element into the treated substrate.

Acknowledgment
This work has been supported in part by Targeted Program of NAS of Ukraine on Plasma Physics and EUROfusion.

References

[1] A. A. F. Tavassoli, E. Diegele, R. Lindau, et al., J. Nucl. Materials **455** (2014) 269-276.
[2] N. Baluc, Plasma Phys. Control. Fusion **48** (2006) B165-B177.
[3] J. W. Coenen, S. Antusch, M. Aumann, et al., Physica Scripta **T167** (2016) 014002.
[4] K. Sugiyama , K. Schmid, W. Jacob, Nuclear Materials and Energy **8** (2016) 1-7.
[5] V. I. Tereshin, A. N. Bandura, A. M. Bovda, et al., Rev. Sci. Instrum.**73** (2002) 831-833.
[6] V. A. Makhlay, I. E. Garkusha, A. N. Bandura, et al., Eur. Phys. J. D. **54** (2009) 185-188.
[7] V. Tereshin, A. N. Bandura, O. V. Byrka, et al., 2004 Vacuum **73** (2004) 555-560.
[8] O. Byrka, et al., Acta Technica **56** (2011) T362-T372.
[9] S. V. Bazdyreva, I. E. Garkusha, et al., Probl. Atom. Sci. Technol. **1(95)** (2015) 166-169.
[10] I. E. Garkusha , Phys. Scripta **91** (2016) 094001.
[11] V. A. Makhlaj, Probl. Atom. Sci. Technol. **1(83)** (2013) 73-75.
[12] V. A. Makhlaj, I. E. Garkusha, N. N. Aksenov, et. al., Phys. Scripta **T161** (2014) 014040.
[13] D. V. Orlinski, V. S. Voitsenya, K. Yu. Vukolov, Plasma Devices Operations **15** (2007) 33-75.
[14] V. A. Makhlaj, I. E. Garkusha, S. V. Malykhin, et al., Physica Scripta **T138** (2009) 014060.
[15] S. V. Malykhin, I. E. Garkusha, V. A. Makhay, et. al., Functional Mater. **24(1)** (2017) 179-183.
[16] M. A. Krivoglaz, Difraction of x-rays and neutrons in non-ideal crystals ("Naukova dumka", Kiev, 1983) p. 408 - in Russian.
[17] V. A. Makhlaj, N. N. Aksenov, et al., Probl. Atom. Sci. Technol. **6(32)** (2016) 129-131.
[18] I. E. Garkusha, O. V. Byrka, V. V. Chebotarev, et al., Vacuum **58** (2000) 195-201.

Strong electromagnetic pulses generated in high-intensity laser-matter interactions

P Rączka[1*], J-L Dubois[2], S Hulin[2], M Rosiński[1], A Zaraś-Szydłowska[1] and J Badziak[1]

[1]Institute of Plasma Physics and Laser Microfusion, 01-497 Warsaw, Poland
[2]CELIA, University of Bordeaux-CNRS-CEA, 33405 Talence cédex, France

[*]Corresponding author: piotr.raczka@ifpilm.pl

Abstract. Results are reported of an experiment performed at the Eclipse laser facility in CELIA, Bordeaux, on the generation of strong electromagnetic pulses. Measurements were performed of the target neutralization current, the total target charge and the tangential component of the magnetic field for the laser energies ranging from 45 mJ to 92 mJ with the pulse duration approximately 40 fs, and for the pulse durations ranging from 39 fs to 1000 fs, with the laser energy approximately 90 mJ. It was found that the values obtained for thick (mm scale) Cu targets are visibly higher than values reported in previous experiments, which is argued to be a manifestation of a strong dependence of the target electric polarization process on the laser contrast and hence on the amount of preplasma. It was also found that values obtained for thin (μm scale) Al foils were visibly higher than values for thick Cu targets, especially for pulse durations longer than 100 fs. The correlations between the total target charge versus the maximum value of the target neutralization current, and the maximum value of the tangential component of the magnetic field versus the total target charge were analysed. They were found to be in very good agreement with correlations seen in data from previous experiments, which provides a good consistency check on our experimental procedures.

1. Introduction

The laser-target interaction at high power and high laser intensity may result in the emission of strong electromagnetic pulses (EMP) with frequencies in the range of tens of MHz to few GHz, as observed already in the seventies [1]. Such pulses may affect the data collection and disturb electronic devices used inside and outside of the experimental chamber, thus making it difficult to realise experiments in a safe and reliable way. With the advent of the petawatt power and MJ energy laser facilities a systematic effort was made to study the EMP phenomenon [2-6]. However, a full quantitative understanding has not been obtained and the problem continues to attract attention [7-14]. In particular, in a series of recent articles [7-9] authors presented a first detailed study of the electric polarization of the target and the resulting neutralization current as a source of EMP. They report on experiments at the fs Eclipse laser facility in CELIA, Bordeaux, in which they studied electric polarization of thick (mm scale) targets. It was found in particular that the charge generated on those targets and the maximum value of the neutralization current are directly proportional to the laser energy at fixed pulse duration and that these quantities are relatively insensitive to the pulse duration at fixed laser energy [7]. In [8] the results on the relation between the target charge and the maximum value of the EMP magnetic field were presented. It is of great interest to extend this analysis to the case of very thin (micrometer scale) foils, such as are commonly used in the laser-driven ion acceleration experiments. To this end a dedicated experiment was performed at the Eclipse facility by our group. Results of this experiment had been described in detail in [15]. In this note we summarize some of those results and compare them in some detail with the results of previous EMP experiments on Eclipse [7,8].

2. Experimental setup

The Eclipse laser at CELIA, Bordeaux, is a Ti:sapphire laser delivering a p-polarized beam with 807 nm central wavelength, focused by an f/5 off-axis parabolic mirror to the focal spot with the FWHM diameter of 10.50 μm. The laser beam was incident on the target along the target normal. The pulse duration was varied in the range 39 fs – 1000 fs, and the laser pulse energy was varied in the range 45 mJ – 95 mJ on target. The ns pedestal due to the amplified spontaneous emission (ASE) was found to be of 5 ns duration. The laser contrast for the majority of shots discussed in this note was 2×10^{-6} at 150 ps before the peak, except for 8×10^{-6} for the shots at pure Al foils and only 7×10^{-5} for some of the initial shots at the thick Cu target. We used the Sophia experimental chamber.

Targets used in our experiment were designed so as to allow for a straightforward comparison between the EMP generated from thin and thick targets. A target holder in the form of a brass ring 14.0 mm in diameter, mounted on a thin brass wire stalk 28.0 mm long was used. Inside the brass ring the following targets were placed: (a) a massive Cu "pill" 10.1 mm in diameter and 1.0 mm thick; (b) an Al foil 6.0 μm thick, with 0.3 μm layer of polystyrene on the rear side, further denoted as "AlPs", pasted on the rear side of a Cu pill 10.1 mm in diameter and 1.0 mm thick, with 10 holes 1 mm in diameter for shots; (c) a pure Al 6.0 μm foil, placed between two Cu pills, each 0.50 mm thick, with 10 holes for shots. The target holder was mounted on a thin square Al plate (160.5 mm×160.5 mm, 4.2 mm thick) that was electrically grounded, to ensure the presence of a mirror charge and the dominance of the dipole component in the radiation pattern.

3. Neutralization current and the total target charge

In order to measure the target neutralization current the target holder stalk was directly connected via a coaxial cable to a high-bandwidth oscilloscope Teledyne LeCroy SDA 760Zi-A (6 GHz bandwidth, 20 GSa/s). The curves representing the temporal dependence of the target neutralization current were very similar for all three types of targets considered here, with a spike followed by few oscillations around the zero point and the timescale of ~1 ns. The net target charge was obtained by integrating over the current spike. The plot of the target charge as a function of the laser energy on target, for the pulse duration of approximately 40 fs, is shown in Fig. 1, together with the data obtained in previous experiments on Eclipse [7,8]. For parameters where three or more consistent data points were obtained in this experiment averages are shown with an error estimate, otherwise we just show a scatter plot.

We see that the values of the target charge obtained for the 1 mm Cu target are visibly higher than the values reported in [7] and [8] for 3 mm Cu target. For example, in [7] the value of (12±1) nC was obtained at 88 mJ and 50 fs, whereas we find (27±2) nC for 85 mJ and 39 fs and even (35±2) nC for 94 mJ and 55 fs. This is puzzling at first glance because both targets are thick enough to stop nearly all fast electrons generated on the front side of the target and in principle should give the same results for the target charge. The only other difference between these measurements is the laser contrast. Indeed, in the previous EMP campaigns on Eclipse the duration of the ASE pedestal was 2 ns and the contrast was 1×10^{-7}, whereas in our experiment the ASE pedestal was 5 ns and the contrast for the data points indicated in Fig. 1 by black dots was 2×10^{-6}. Thus it is our hypothesis that the difference in the target charge is a manifestation of a strong sensitivity of the target charge to the laser contrast and hence to the scale of the preplasma. This conjecture is supported by the fact that in the few shots made at 94 mJ and 55 fs that gave the extremely high value of the target charge the laser contrast was only 7×10^{-5}. Apart from this data point all our data for the 1 mm Cu target is in very good agreement with the direct proportionality of the target charge to the pulse energy, similarly to what was observed in [7] for the 3 mm Cu target, except that the slope is higher.

Turning to thin foil targets we see that the charge they generate is visibly higher than that generated from the 1 mm Cu target. For example, for the AlPs target we measure (32±2) nC at 86 mJ and 41 fs, which is 20% higher than the charge obtained for the 1 mm Cu target at similar laser parameters. The charge measured for the pure Al target appears to be higher by an even larger percentage, 50% or more, although this might be partially due to worse laser contrast (8×10^{-6}).

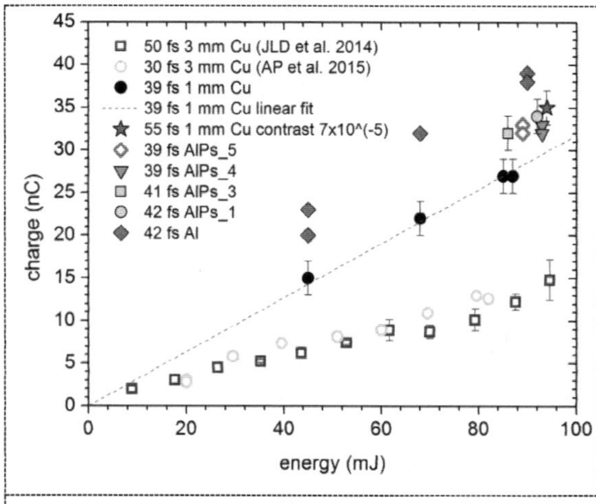

Figure 1. The total target charge as a function of the laser pulse energy on target, for the pulse duration close to 40 fs, as obtained in this experiment, compared with the data from previous experiments [7,8].

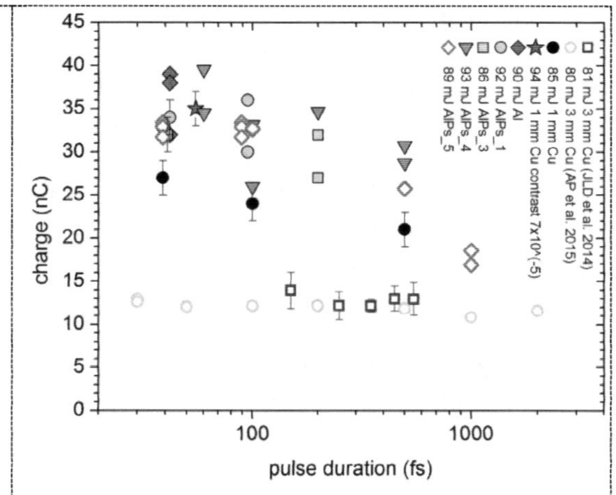

Figure 2. The target charge as a function of the pulse duration, for the pulse energy close to 90 mJ, as obtained in this experiment, compared with the data from previous experiments [7,8].

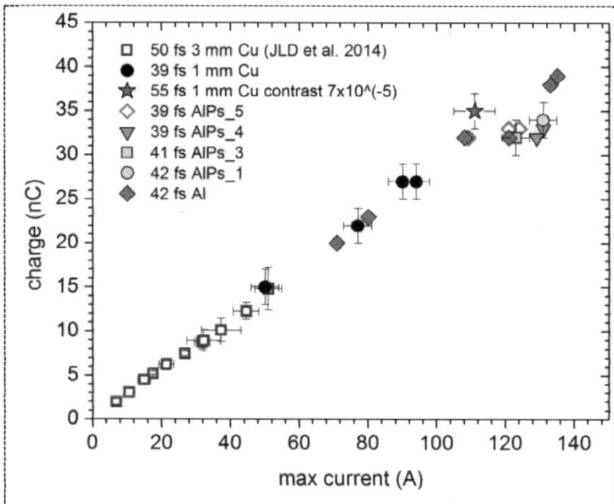

Figure 3. The correlation plot between the total target charge and the maximum value of the target neutralization current, for the data taken in the laser energy scan, for the pulse duration close to 40 fs, as obtained in this experiment, compared with the data from previous experiments [7,8].

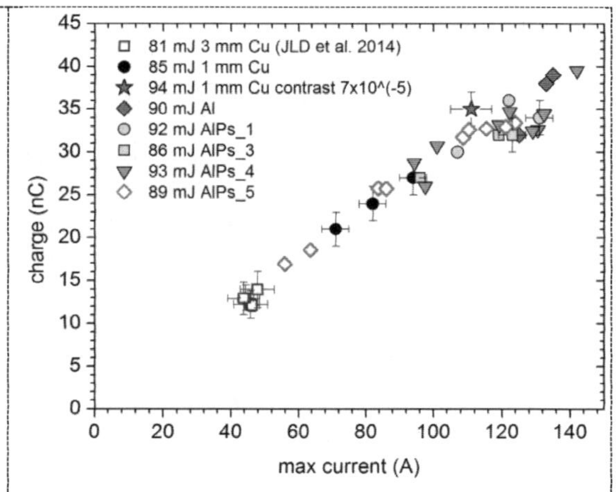

Figure 4. The correlation plot between the total target charge and the maximum value of the target neutralization current, for the data taken in the pulse duration scan, for the pulse energy close to 90 mJ, as obtained in this experiment, compared with the data from previous experiments [7,8].

In Fig. 2 we show the total target charge as a function of the pulse duration, for the pulse energy close to 90 mJ, together with the data obtained in previous experiments on Eclipse [7,8]. The following remarks can be made about the data from our experiment shown there: (1) for all values of the pulse duration the data from 1 mm Cu targets are visibly higher that the values reported in earlier experiments [7,8] (our hypothesis is that this is due to the difference in the laser contrast, as explained above); (2) the data for 1 mm Cu target show substantial dependence on the laser pulse duration, whereas data from earlier experiments are practically insensitive to the pulse duration (again, we interpret this qualitative difference as an interesting manifestation of the sensitivity to the laser contrast and the presence of a preplasma); (3) similarly as in the laser energy scan the charge obtained with thin targets is visibly higher that the charge generated with the thick targets.

As was mentioned above, the temporal profile of the neutralization current was nearly identical in all measurements, except for the maximum value of the current [15]. In such a case there should be a strong correlation between the values of the target charge and the maximum values of the return current. In Fig. 3 we show such correlation for the data from the pulse energy scan, and in Fig. 4 we display data obtained in the pulse duration scan (plots showing the data for the maximum values of the return current in these scans had been given in [15]). On both figures we see that the correlation is indeed very strong. This fact provides an important consistency check for our analysis. Furthermore, we see that the trend in the data obtained in [7] is in excellent agreement with the trend in the data from the present experiment, which provides a very strong argument that the difference between the experiments discussed above is not a result of some error but indeed a consequence of different physical conditions under which the measurements were made (i.e. the laser contrast). Finally, we note that several points in Fig. 3 and Fig. 4 corresponding to shots at the AlPs target lie below the line traced by other data points, which may indicate that there may be additional effects in the target charge generation and redistribution processes in such targets related to the presence of a Ps layer at the rear of the target.

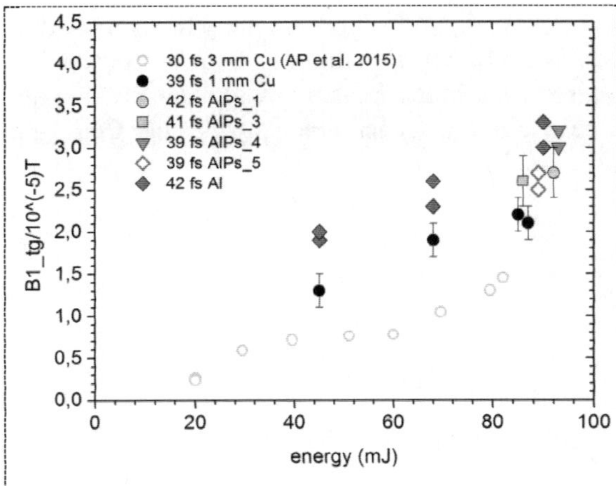

Figure 5. The maximum value of the tangential component of B as a function of the laser energy, for the pulse duration close to 40 fs, as obtained in this experiment, compared with the data from previous experiment [8].

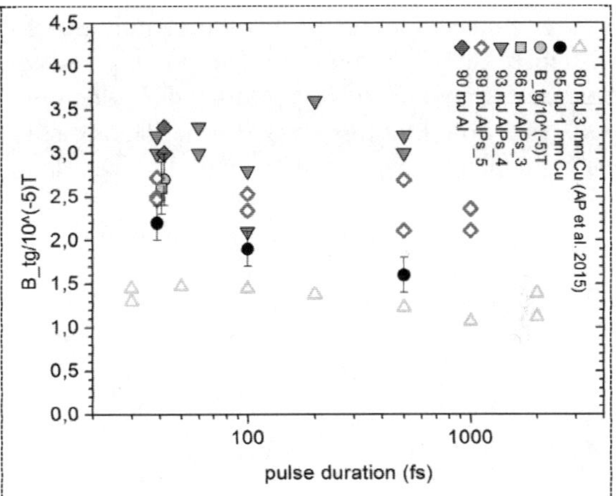

Figure 6. The maximum value of the tangential component of B as a function of the pulse duration, for the laser energy close to 90 mJ, as obtained in this experiment, compared with the data from previous experiments [8].

4. Measurement of the EMP magnetic field

In order to characterize the EMP signal generated in laser-target interaction the magnetic field induction was measured inside the experimental chamber using a radiation-hardened Prodyn RB-230 B-dot probe, connected to a Prodyn BIB-100G balun. The Teledyne LeCroy oscilloscope connected via a 2 m long coaxial cable was used to collect the data. The probe was placed in a point located 214 mm behind the target, 41 mm above the target, and 55 mm to the right of the target, and set to measure the tangential component of the B field (relative to the natural cylindrical coordinate system in which the axis coincides with the vertical axis of the target holder). The data on dB_{tg}/dt was processed by applying to the discrete

IOP Conf. Series: Journal of Physics: Conf. Series **959** (2018) 012005 doi:10.1088/1742-6596/959/1/012005

Fourier transform a band pass filter with 400 MHz lower cutoff and 7.5 GHz upper cutoff and correcting for the frequency-dependent attenuation of the cable, and then integrating to obtain $B_{tg}(t)$. The resulting signal has the form of a strong initial spike followed by decaying oscillations lasting approximately 200 ns, interspersed by what appears to be reflections of the original spike from the chamber walls and the conductive elements inside the experimental chamber. A detailed description of $B_{tg}(t)$ may be found in [15].

From the point of view of the EMP studies it is the initial spike that is most interesting, as it generally corresponds to the highest values of the field measured in the whole signal. Furthermore, it is also a signal originating directly from the laser-target interaction, undisturbed by any reflections or contamination by the chamber eigenmodes, and hence it is of more fundamental character. In order to characterize the magnitude of EMP a maximum value of the B_{tg} was determined for each shot. In Fig. 5 we show the maximum value of B_{th} as a function of the laser energy, with pulse duration kept close to 40 fs, and in Fig. 6 we show the same quantity as a function of the laser pulse duration, with pulse energy kept close to 90 mJ. On both figures the data obtained in [8] is also indicated for comparison. It should be noted that there is a normalization error in Fig. 4 of [8]: the values of B_{tg} indicated in this figure have to be divided by a factor 2π.

The following general statements can be made about the maximum values of B_{tg} recorded in our experiment: (1) the values for the 1 mm Cu target are visibly higher than values reported in [8] for the 3 mm Cu target (after correcting it for the omitted factor of 2π); (2) the values obtained for thin foil targets are systematically higher than values obtained for 1 mm Cu target. Interestingly, Fig. 5 looks quite similar to Fig. 1. In principle the same could be said about the Fig. 6 and Fig. 2, with an important exception of the measurements for thin targets at pulse durations longer than 100 fs, in which case we noticed a surprising effect that the magnitude of the initial spike retains approximately the same value with increasing pulse duration, despite a greatly reduced overall EMP signature. It should be noted that the maximum value of B_{tg} measured for the 3 mm Cu target and longer pulses is also insensitive to the pulse duration, but apparently for a different reason, because in our measurements for 1 mm Cu target we see a clear evolution of B_{tg} with the increasing pulse duration.

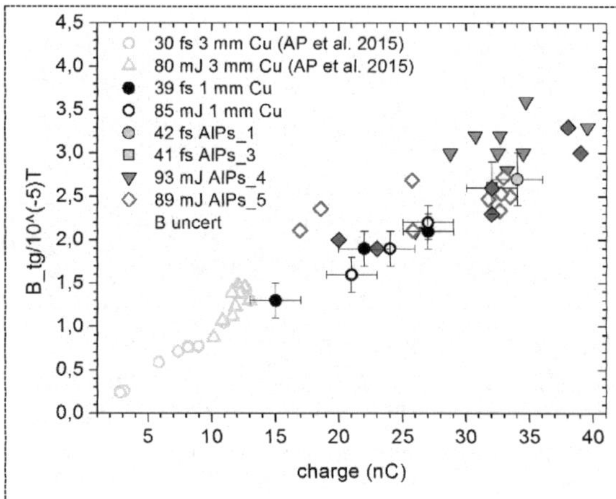

Figure 7. The correlation plot between the maximum value of the tangential component of the magnetic field and the total target charge, as obtained in this experiment in the laser energy scan at 40 fs pulse duration, and a pulse duration scan at 90 mJ laser energy, compared with the data reported in [8] for the 3 mm Cu target.

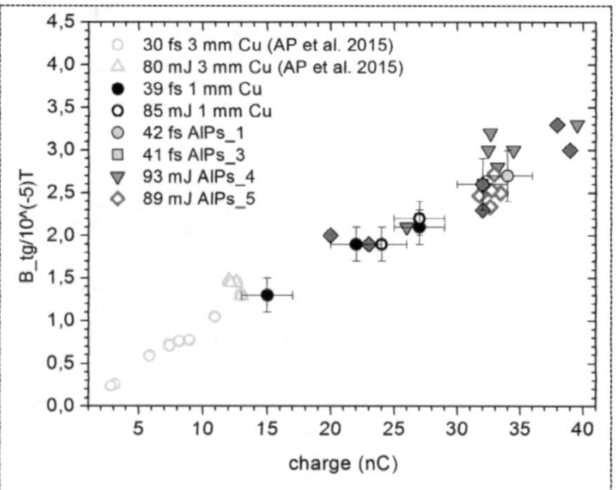

Figure 8. The same as in Fig. 7, excluding the data points corresponding to pulse duration longer than 100 fs.

The fact that Fig. 5 is quite similar to Fig. 1 suggests that it may be useful to study the correlation between the maximum value of B_{tg} and the total target charge. Such a correlation is shown in Fig. 7, where all the data shown in previous figures for the laser energy scan and the pulse duration scan are included. It is interesting to compare this figure with Fig. 4 in [8]. We see that in general a relatively strong correlation is found. As we see, a nonlinear trend that seemed to be indicated by the data shown in [8] is not supported at higher values of the target charge. It is nevertheless evident that the scatter is much bigger than for example in Fig. 3 and Fig. 4 for the correlation between the charge and the maximum value of the current. To check whether this may be a consequence of an unusual behavior of the initial spike at larger values of the laser pulse we show in Fig. 8 the data from Fig. 7 with all points corresponding to the pulse duration longer than 100 fs removed from the plot. We see that the correlation in the data set restricted in this way is much stronger, which supports the hypothesis that for longer pulse durations other mechanisms of EMP generation might become relevant, apart from a straightforward neutralization of the target charge. Furthermore, the data from our experiment seems to be in excellent agreement with the data reported in [8], after correcting it for the omitted factor of 2π. This provides yet another important cross check on the validity of our measurements and confirms high quality of our data analysis.

5. Conclusions

Summarizing, we performed measurements of the target neutralization current, the total target charge and the tangential component of the magnetic field of the resulting electromagnetic pulse for the laser energies ranging from 45 mJ to 92 mJ, with to pulse duration approximately 40 fs, and for the pulse durations ranging from 39 fs to 1000 fs, with the laser energy approximately 90 mJ. We found that values obtained for the 1 mm Cu target are visibly higher than values obtained in previous experiments [7,8] for the 3 mm Cu target, which we argue is a manifestation of a strong dependence on the laser contrast and hence on the amount of preplasma (the sensitivity of EMP to the presence of preplasma was noted in [16]). We also found that measurements for 6 μm Al foils give visibly higher results than for the thick Cu targets, which is especially true for pulse durations longer than 100 fs. Looking at plots of the target charge versus the maximum value of the neutralization current and the maximum value of the tangential component of the magnetic field versus the total target charge we find a very high degree of correlation, which provides an important check of the validity of our data collection and consistency of our data analysis. We also find very good agreement with the data obtained in previous experiments [7,8], after an error in normalization of the magnetic field reported in [8] is corrected.

Acknowledgements

The authors are grateful to V. Tikhonchuk for discussion and support and A. Poyé for comments on the results presented in [8]. This research is supported by the Polish National Science Centre grant Harmonia 2014/14/M/ST7/00024. Access to the Eclipse facility was made possible by the support obtained from LASERLAB-EUROPE, European Union's Horizon 2020 research and innovation programme, under the grant agreement no. 654148, project CNRS-CELIA002294. Our collaboration has profited greatly from the intellectual environment created by the COST Action MP1208.

References

[1] Pearlman J S and Dahlbacka G H 1977 *Appl. Phys. Lett.* **31** 414.
[2] Mead M J, Neely D, Gauoin J, Heathcote R and Patel P 2004 *Rev. Sci. Instrum.* **75** 4225.
[3] Raimbourg J 2004 Rev. Sci. Instr. **75** 4234.
[4] Stoeckl C et al. 2006 *Rev. Sci. Instrum.* **77** 10F506.
[5] Bourgade J L et al. 2008 *Rev. Sci. Instrum.* **79**, 10F301.
[6] Brown Jr. C G, Throop A, Eder D and Kimbrough J, *J. Phys. Conf. Ser.* **112** 032025.
[7] Dubois J-L et al. 2014 *Phys. Rev.* E **89** 013102.
[8] Poyé A et al. 2015 *Phys Rev.* E **91** 043106.

[9] Poyé A et al. 2015 *Phys Rev.* E **92** 043107.
[10] Krása J et al. 2015 *Laser Part. Beams* **33** 601.
[11] Consoli F. et al. 2016 *Sci. Rep.* **6** 27889.
[12] Robinson T S et al. 2017 *Sci. Rep.* **7** 983.
[13] Krása J et al. 2017 *Plasma Phys. Control. Fusion* **59** 065007.
[14] De Marco M et al. 2017 *Phys. Plasmas* **24** 083103.
[15] Rączka P, Dubois J-L, Hulin S, Tikhonchuk V, Rosiński M, Zaraś-Szydłowska A and Badziak J 2017 *Laser Part. Beams*, published online on 2 Nov. 2017 (https://doi.org/10.1017/S026303461700074X)
[16] Varma S et al. 2014 *Opt. Eng.* **53** 051515

Analysis of optical spectra from steel samples exposed to pulsed plasma streams

A Marchenko[1], I Garkusha[1], V Makhlai[1], E Skladnik-Sadowska[2], R Kwiatkowski[2], K Nowakowska-Langier[2], M J Sadowski[2-3], M Kubkowska [3]

[1]Institute of Plasma Physics of the NSC KIPT, Akademicheskaya St.1, 61108 Kharkov, Ukraine
[2]National Centre for Nuclear Research, A. Soltana str. 7, 05-400 Otwock-Swierk, Poland
[3]Institute of Plasma Physics and Laser Microfusion, Hery str. 23, 01-497 Warsaw, Poland

Abstract. In the reported studies the attention was focused on an analysis of optical spectra of the visible radiation emitted from stainless steel (SS) samples exposed to intense pulses from a ND:YAG laser or pulsed plasma streams produced by a Rod Plasma Injector (RPI) IBIS. Operational regimes of the RPI IBIS facility, which can generate intense deuterium-plasma streams, were studied and corresponding energy fluxes were determined. Optical emission spectra from the laser- or plasma-irradiated SS samples were recorded at different energies supplied. It was proved that intensities of the emitted spectral lines of Fe I and Cr I species depend strongly on deuterium-plasma energy deposited upon the irradiated target.

1. Introduction

The austenitic stainless-steel (SS) is a very good material to construct different parts of plasma facilities, e.g. vacuum chambers, in-vessel components, diagnostic ports, etc. Several years ago, it was recommended that the main constructional material of the ITER facility should be of the SS type [1]. Therefore, it was of interest to investigate behaviour of SS samples under irradiation by intense plasma streams, and to study damages caused by plasma containing fast electrons and ions (particularly fast deuterons and protons). The main question was to investigate which SS type is the most resistant to erosion induced by plasma impacts. It was known that the optical emission spectroscopy can deliver important information about dynamics and parameters of the investigated plasma streams. Such data are of primary importance for research on the interaction of plasma streams with various targets. Therefore, new experiments have been undertaken for different SS types, using intense laser pulses and pulsed plasma streams.

2. Experimental setup and diagnostics

Detailed investigations were carried out for two types of SS samples. A chemical composition of them is given in a Table 1.

Table 1. Chemical composition of the investigated SS types.

Main elements [%]	SW7	S316/316L
Cr	3.5-4.5	16-18
Ni	0.4	10-14
Mo	4.5-5.5	2-3
W	6-7	-
Fe	basic	basic

IOP Conf. Series: Journal of Physics: Conf. Series **959** (2018) 012006 doi:10.1088/1742-6596/959/1/012006

Irradiation tests of the chosen SS samples of dimensions 3 x 2 x 0.2 cm^3were performed within two different experimental facilities. In the first experiments the samples were exposed to laser pulses only, and during the main experiments they were irradiated by intense deuterium plasma streams.

During the first experimental series the SS samples were placed at the centre of the vacuum chamber and irradiated by intense pulses from a Nd:YAG laser of the LOTIS TII type, which operated at λ_L= 1064 nm and delivered 3 ns-pulses of energy from 0.1 J to 0.5 J.The laser beam was introduced at an angle of 30° to the target normal and focused (by means of achromatic lenseswith f = 42 cm) to the spot of 1 mm in diameter. Optical observations were carried out at an angle of 30^0 to the target normal,through a quartz collimator and an optical cable. Spectroscopic measurements of the visible radiation, which was emitted from plasma produced during the interaction of a laser beam with the target, were performed by means of a Mechelle®900 optical spectrometer. It was able to record optical spectra in the wavelength range from about 300 nm to 1100 nm, at expositions varied from 100 ns up to 50 ms.

The main experimental studies have been performed within the RPI-IBIS facility equipped with molybdenum (Mo) electrodes [2-4], which were powered from a condenser bank charged to U_0 = 28 kV. The peak discharge current amounted to I_{max} = 550 kA. The operational mode of the device was varied by changes of a time delay between the gas puffing and the application of a high-voltage pulse. In order to record optical emission spectra in the RPI-IBIS facility the use was made of the same Mechelle®900 spectrometer, which observed the irradiated sample through a quartz collimator and an optical cable. The positioning of the SS target, which was placed at a distance of about 20 cm from the electrodes outlet, as well as typical discharge traces are shown in Figure 1.

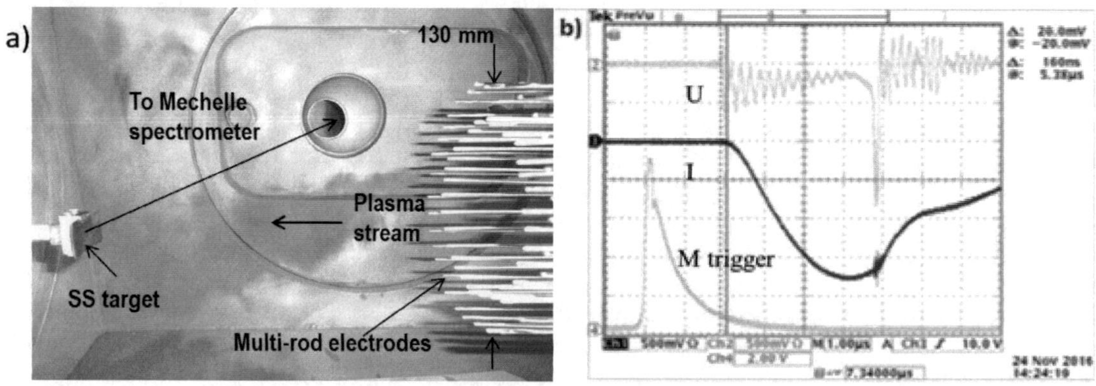

Figure 1. Location of a SS target inside the RPI-IBIS vacuum chamber (a), and a temporal correlation of the voltage $U_d(t)$ and current $I_d(t)$ traces with a pulse used to trigger the Mechelle-900 spectrometer (b).

3. Experimental results and discussion

3.1. Results of the irradiation of the SS samples by laser pulses

During the preliminary experiments the investigated SS samples were exposed to focused Nd:YAG laser pulses of the same duration (3 ns), but different energy which was equal to E_L = 0.15 J, 0.2 J, 0.3 J and 0.44 J. Since the samples were irradiated under vacuum conditions, at the background pressure p_0 = 6.8 x 10^{-5} mbar, the radiation emitted from laser-produced plasma had an optical spectrum composed of the spectral lines originated only from components of the irradiated target-material. There were no lines from other materials and impurities which are often observed during experiments with pulsed plasma streams.

The visible radiation emitted from plasma produced in the near-target region, due to erosion of laser-irradiated samples, was recorded by means the Mechelle®900 spectrometer described above. Some examples of the recorded optical spectra are presented in Figure 2.

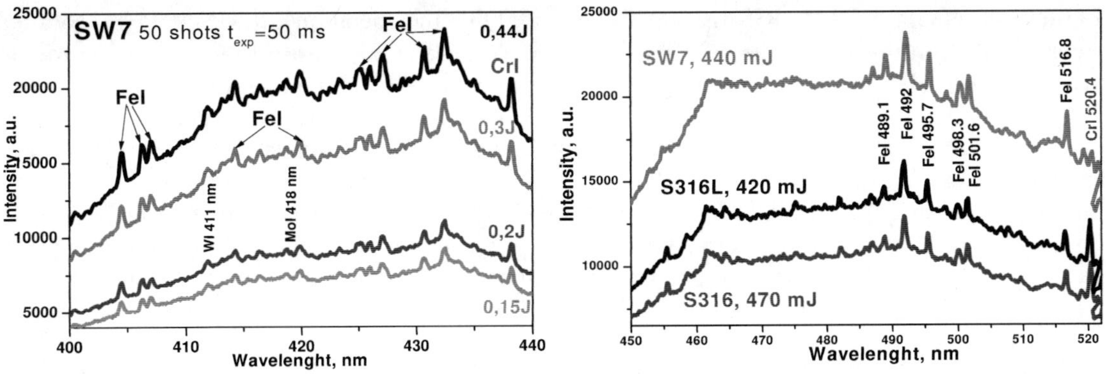

Figure 2. Part of the optical spectrum from plasma produced from the SW7 samples irradiated by laser pulses of different energy (a) and another part of the optical spectra obtained from different SS samples irradiated by laser pulse of similar energy (b).

The optical spectra presented in Figure 2a, as recorded for plasma created by surface erosion of the SW7 samples, showed that the near-surface plasma production and intensity of its radiation depended considerably on laser pulse energy. There were identified distinct spectral lines originated from Fe I species (because iron was the main component of the SS samples), as well as intense lines of W I – 411 nm, Mo I – 418 nm, and Cr I – 432 nm (originated from other sample-components). As expected, intensities of all the mentioned spectral lines increased with an increase in laser pulse energy.

A comparison of the optical spectra obtained from SW7, S316 and S316L samples exposed to laser pulses of comparable energy, as presented in Figure 2b, showed that intensity of the plasma radiation from the S316L sample was about 1.5 times lower than that from the SW7 sample. In the case of the S316 sample the optical emission was still lower in spite of the fact that laser-pulse energy was about 10% higher. Hence, one could conclude that the S316 sample has higher resistance to erosion induced by a laser impact. Therefore, it was reasonable to check whether this effect occurs also when the chosen SS samples are exposed to pulsed plasma streams impinging upon the target surface.

3.2. *Results of irradiation of the SS samples by pulsed plasma streams*

The main experiments on the erosion of the SS samples were performed within the RPI-IBIS facility described above. The optimal operational mode of this device was found on the basis of the optical emission spectra presented in Figure 3.

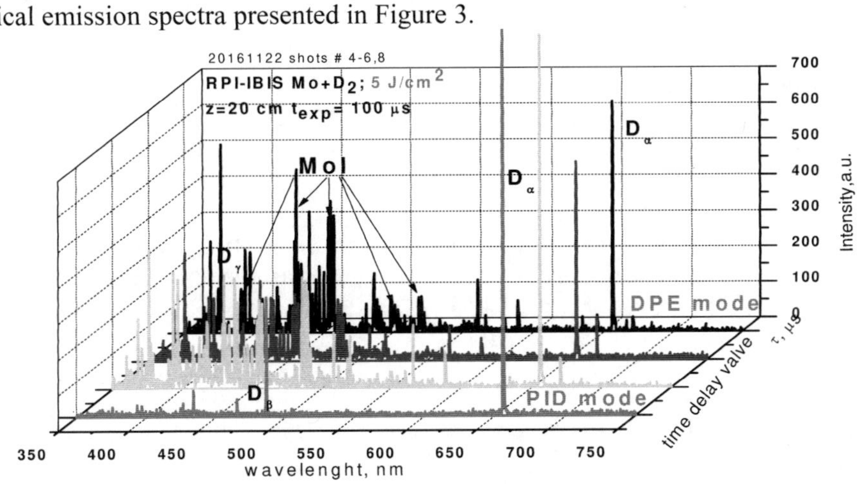

Figure 3. Optical emission spectra of pulsed plasma streams generated within the RPI-IBIS facility, which were recorded for different operational modes.

IOP Conf. Series: Journal of Physics: Conf. Series **959** (2018) 012006 doi:10.1088/1742-6596/959/1/012006

Earlier experiments with the RPI-IBIS facility showed that the operational mode could be modified by a change of a time delay τ of the discharge ignition in relation to the gas injection [2-4]. In order to expose the SS samples to clean deuterium plasma streams a so-called PID mode was chosen at $\tau = 215$ μs, and plasma energy-flux density equal to 8 J/cm^2. On the basis of the optical spectra, analyzing a linear Stark effect of D-lines, the Voigt contour fitting and the know formula $\Delta\lambda_{1/2} \approx n_e^{2/3}$ [5], it was estimated that the electron density of a freely-propagating plasma stream was (1.1-2.2) x 10^{15} cm^{-3}.

During the main experimental session the chosen S316 sample was irradiated in the RPI-IBIS facility, and the optical spectrum from near-surface plasma was recorded, as presented in Figure 4a. For a comparison a SW7 sample was also irradiated under the identical experimental conditions, and the corresponding spectrum was obtained, as shown in Figure 4b.

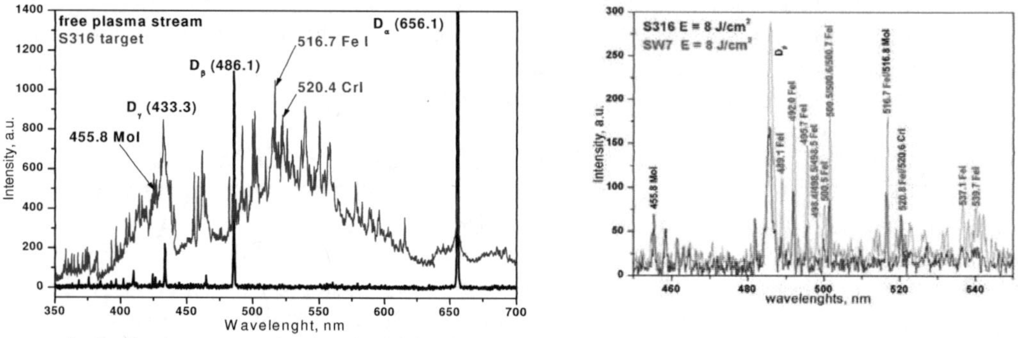

Figure 4. Optical spectra obtained within the RPI IBIS facility without and with the S316 target (a), and a comparison of the spectra recorded for the irradiated S316 and SW7 samples (b).

From the spectra in Figure 4a one could easily see that during interactions of a deuterium plasma stream with the S316 target there were emitted many spectral lines of Mo I, Fe I and Cr I, which evidently originated from the eroded material. In that case the electron density of near-surface plasma (integrated over the whole discharge period) amounted to (3-7) x 10^{15} cm^{-3}. During discharges when the spectrometer exposition time was shorter (e.g. 5 μs) the electron density amounted 5 x 10^{16} cm^{-3}.

A comparison of the spectra shown in Figure 4b confirmed that intensities of the observed spectral lines from the S316 sample were about twice lower than those from the SW7 target. It should here be noted that intensity of the Mo I - 455.8 nm line, which originated from the Mo-electrode erosion, remained unchanged for experiments with both types of the SS samples.

4. Summary and conclusions

Interactions of laser pulses and pulsed plasma-streams with different austenitic stainless-steels (SW7 and S316/316L) were investigated. The operational regimes of the RPI-IBIS facility were controlled and corresponding plasma energy flux densities were measured. During irradiations of the chosen samples many spectral lines from iron and other target-components were recorded. The samples of the S316/316L type showed higher resistance to erosion than the SW7 samples.

Therefore, for the construction of plasma facing components it was recommended to use the S316/316L stainless-steel. It was also proved that the RPI-IBIS facility may be applied for research on interactions of plasma streams with materials at power fluxes amounting to 5 MW/cm^2.

References

[1] D. Stork, S.J. Zinkle, Nucl. Fusion 57 (2017) 092001 (9pp).
[2] M.J. Sadowski, J. Baranowski, et al., Appl. Surface Science 238, Issues 1-4 (2004) 433-437.
[3] J. Piekoszewski, J. Stanislawski, et al., Czech. J. Phys. 54, Suppl. C (2004) C217-C222.
[4] E. Skladnik-Sadowska, K. Czaus, et al., Nukleonika57, No 2(2012) 193–196.
[5] Plasma Diagnostic Techniques, Edit. R.H. Huddlestone, S.L. Leonard (Academic Press, 1965).

Study of Plasma Behavior during ECRH Injection in the GAMMA 10 SMBI Experiments

Md. Maidul Islam[1*], Yousuke Nakashima[1], Shinji Kobayashi[2], Nobuhiro Nishino[3], Kazuya Ichimura[4], Takaaki Iijima[1], Md. Shahinul Islam[1], Takayuki Yokodo[1], Guanyi Lee[1], Tsubasa Yoshimoto[1], Sotaro Yamashita[1], Masayuki Yoshikawa[1], Junko Kohagura[1], Mafumi Hirata[1], Ryutaro Minami[1], Tsuyoshi Kariya[1], Ryuya Ikezoe[1], Makoto Ichimura[1], Mizuki Sakamoto[1] and Tsuyoshi Imai[1]

[1]Plasma Research Center, University of Tsukuba, Tsukuba, Ibaraki 305-8577, Japan
[2]Institute of Advanced Energy, Kyoto University, Gokasho, Uji 611-0011, Japan
[3]Graduate School of Engineering, Hiroshima University, Hiroshima 739-8527, Japan
[4]Electrical and Electronic Engineering Dept., Kobe University, Kobe-657-8501, Japan

e-mail: maidul@prc.tsukuba.ac.jp, maidulphysics@yahoo.com

Abstract. Establishment of fueling system is one of the critical issues for the future fusion reactors. Fueling experiment supersonic molecular beam injection (SMBI) have been carried out in the central-cell of GAMMA 10. In GAMMA 10, electron cyclotron resonance heating (ECRH) is used at plug/barrier-cells for the formation of the axial confining potential. Recently, ECRH was applied during SMBI to plug the loss particles and increased the plasma density in the central-cell compared to without ECRH. This result suggests that the particles are confined during SMBI due to the injection of ECRH at plug/barrier-cells in GAMMA 10.

Keywords: GAMMA 10, supersonic molecular beam injection, high density plasma formation, ECRH.

1. Introduction

Optimization of fueling control in the magnetically confined plasmas enables the profile control of the core plasma density as well as reduction in neutral particles in the peripheral area. For example, ice-pellet injection [1, 2] is effective in increasing the density in the plasma core region, while gas puffing [3, 4] is suitable to control the density in the edge region including the divertor. The ice-pellet injection is characterized by a high fueling efficiency however, it is successful in large devices. Supersonic molecular beam injection (SMBI) developed by L. Yao et al. [5], is suitable for the control of edge-localized-mode in ITER-like tokamaks [6]. SMBI gives high-speed and highly-directive gas injection system which can inject neutral particles deeper into the core-plasma. In order to understand the mechanism of the high fueling efficiency of SMBI, many experiments have been carried out in many devices, such as HL-2A tokamak [7, 8], Heliotron J [9, 10], NSTX [11], JT-60U [12], Tore Supra [13], EAST tokamak [14] and Large Helical Device [15].

In order to form high-β plasma in the mirror device, it is an important issue to increase the plasma density in the main plasma region. Recently, fueling experiment by SMBI has been started to increase

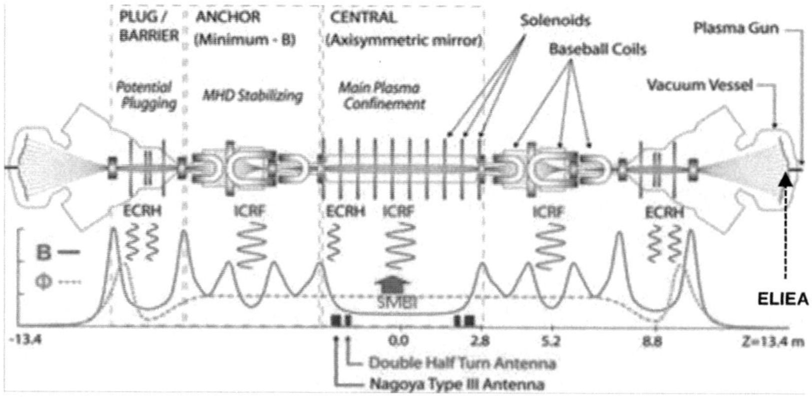

Fig.1 Schematic view of GAMMA 10 and axial profile of magnetic field strength B and potential together with location of SMBI system and ELIEA.

the plasma density in the main plasma confinement region of the GAMMA 10 tandem mirror [16-18]. GAMMA 10 is the world largest tandem mirror device which is axisymmetrized minimum-B anchored with thermal barrier at both mirror ends [19, 20]. In GAMMA 10 SMBI has been applied in the central-cell and has enabled to obtain high density plasmas in the central-cell (NLcc) as compared with the conventional gas puffing. It inputs a large amount of gas in only a short time. The increase in the amount of gas in the central-cell increases the plasma density. Therefore, more ions flow into the end-cells, and the ion-current density is increased during SMBI. Electron cyclotron resonance heating in plug/barrier region (P/B-ECRH) has been applied in GAMMA10 for the formation of the axial confining potential to plug the lost particles and has been observed an increase in the central-cell electron density [21]. In this case, it has been observed that a reduction in the lost-ion flux escaping from the main confining region to the both mirror ends. Therefore, it is important subject to investigate the effect of P/B-ECRH in the SMBI fueled plasmas and its response to the end-loss ions. In this study, we applied the central-cell SMBI in the P/B-ECRH plasmas to study the effect of ECRH to the plasma parameters. The response of the SMBI to the end-loss ions is investigated in the P/B-ECRH plasmas.

2. GAMMA 10 and Experimental setup

A schematic view of GAMMA 10 tandem mirror together with heating is shown in Fig. 1. The total length of GAMMA 10 is 27 m. It consists of central-cell, anchor-cells, plug/barrier-cells and end-cells. The main confinement region of GAMMA 10 is the central-cell (Z=-300 ~ +300 cm). The initial plasma produced by the plasma gun is heated by ICRF waves and confined by magnetic mirror field and electrostatic potential. Two choke coils and ten pancake coils are installed and formed the simple axi-symmetric magnetic mirror configuration in the central-cell. The both ends of central-cell are connected to east/west anchor-cells, which keep the plasma stably. Three baseball coils are installed and form the average minimum-B configuration in each anchor-cell. The outside of anchor-cells are connected to east/west plug/barrier-cells, in which plug potential and barrier potential are produced. East/west end-cells are located to the outside of plug/barrier-cells in which the plasma flow out. The end-cells are at the very end of GAMMA 10 where end-loss ion energy analyzer (ELIEA) is installed to measure the end-loss ion current density. SMBI is very simple system, where high-pressure hydrogen gas is ejected through a fast solenoid valve equipped with a Laval nozzle. In GAMMA 10, SMBI system has been installed in the central-cell at Z=-14.5 cm. In the experiment, the plenum pressure is varied from 0.3 MPa to 2.0 MPa and the pulse width is usually 0.5 ms.

3. Results and discussion

In the experiment, SMBI pulses were injected into the typical plasmas heated by ICRF and ECRH. SMBI pulse is injected at 150 ms and the pulse width is 0.5 ms. The electron line-density is increased

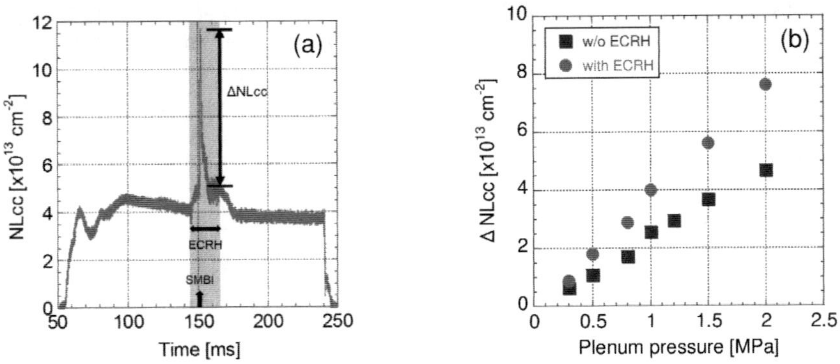

Fig. 2 (a) Temporal behavior of electron line density and (b) comparison of change in electron line density during SMBI with ECRH and without ECRH case.

during SMBI. Figure 2(a) shows the temporal behavior of NLcc during SMBI with P/B-ECRH injection. P/B-ECRH injected at 145 ms and duration is 20 ms and SMBI is injected during ECRH injection. P-ECRH and B-ECRH power are 150 kW and 100 kW, respectively. Figure 2(b) shows the change in electron line-density during SMBI with ECRH and only ICRF heated plasma. From this figure it is observed that in both the ICRF and ICRF/ ECRH plasmas, the change in the electron line-density in the central-cell increases almost linearly to the plenum pressure. From the previous study, it has been obtained that the directivity of the SMBI gas improved when the plenum pressure was up to 1.0 MPa and it saturated more than 1.0 MPa [17]. In such the case, the change in the electron line-density is almost proportional to the change in the Hα intensity whose sightline observes the SMBI gas. These results indicate that the particle source produced by SMBI has a contribution to the increase in the electron density without saturation under the present experimental condition. Note that, in all plenum pressure cases, the change in electron line density is higher in the case of ECRH injection than the ICRF heated plasma. It implies that the particles are confined in the central-cell due to P/B-ECRH injection in the plug/barrier region.

In order to measure the end-loss ion current, ELIEA is installed on both end sides in GAMMA 10. The end loss ion current is measured by using sweep repeller voltage. Figure 3(a) shows that the end-loss

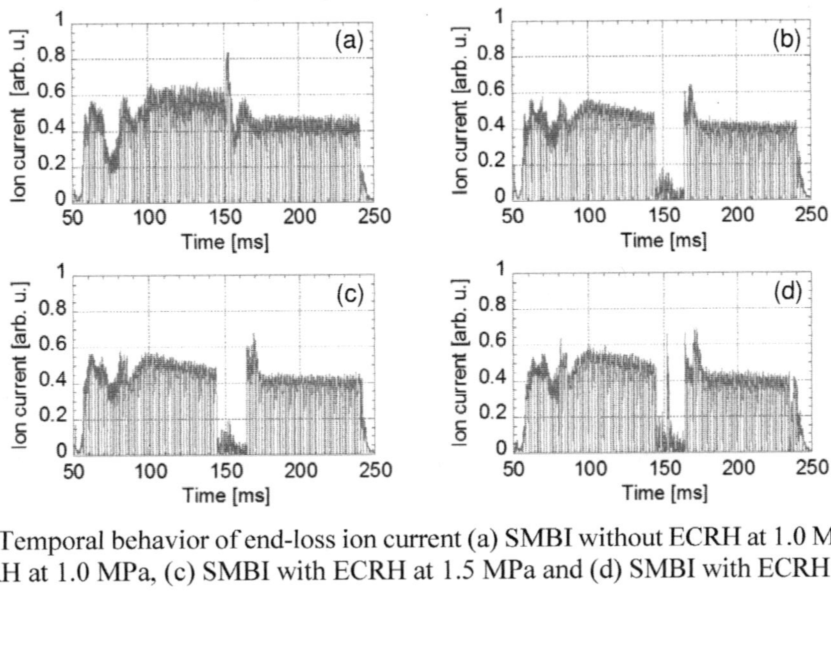

Fig. 3 (a) Temporal behavior of end-loss ion current (a) SMBI without ECRH at 1.0 MPa, (b) SMBI with ECRH at 1.0 MPa, (c) SMBI with ECRH at 1.5 MPa and (d) SMBI with ECRH at 2.0 MPa.

ion current is increased during SMBI at plenum pressure 1.0 MPa. SMBI inputs a large amount of gas within a short period of time. The plasma density increases in the central-cell due to the increase in the amount of gas. Therefore, more ion flow into the end-cells, and the ion current density is increased during SMBI. In order to plug the particles escaped from central-cell to end-cells, P/B-ECRH is applied to ICRF heated plasma. Therefore, the end-loss ion current is reduced and increases the plasma density in the central-cell during SMBI at plenum pressure 1.0 MPa as shown in figure 3(b). It might be due to the formation of plug and thermal barrier potential by P/B-ECRH in the plug/barrier region. However, SMBI at high plenum pressure (~2.0 MPa), the P/B-ECRH could not suppressed the end loss current remarkably. At high plenum pressure the amount of gas is high therefore the end-loss ion current density is high which could not be suppressed by this low power of ECRH system.

4. Summary

SMBI experiment has been performed in the central-cell during ECRH injection. Experimental results implies that the electron line-density is increased during SMBI with ECRH due to the confinement of particles in the central-cell. The end-loss ion current is suppressed at low plenum pressure by the production of plug/barrier potential in the plug/barrier region and increases the plasma density in the central-cell. In future, we will apply high ECRH power to suppress the end-loss ion current at high plenum pressure and increase the plasma density in the central-cell.

Acknowledgments

This study was supported by the bilateral collaboration research program in the University of Tsukuba, Kyoto University and Hiroshima University (NIFS12KUGM068, NIFS12KUHL050 and NIFS12KUGM065). The author sincerely thanks the members of the GAMMA 10 group for their collaboration on the experiments.

References

[1] Lang P T et al 1996 Nucl. Fusion **36** 1531.
[2] Sakamoto R et al 2001 Nucl. Fusion **41** 381.
[3] Miyazawa J et al 2004 et al Nucl. Fusion **44** 154.
[4] Nakano T et al 2008 Nucl. Fusion **48** 085002.
[5] Yao L et al 2007 Nucl. Fusion **47** 1399.
[6] Kim J et al 2012 Nucl. Fusion **52** 114011
[7] Yu D L et al 2010 Nucl. Fusion **50** 035009.
[8] Yu D L et al 2012 Nucl. Fusion **52** 082001.
[9] Zang L et al 2014 Phys. Plasmas **21** 042308.
[10] Mizuuchi T et al 2010 Contrib. Plasma Phys. **50** 639.
[11] Soukhanovskii V A et al 2004 Rev. Sci Instrum. **75** 4320.
[12] Takenaga H et al 2010 Nucl. Fusion **50** 115003.
[13] Pégourié B et al 2003 J. Nucl. Mater. **313-316** 539.
[14] Zheng X et al 2013 Plasma Phys. Control. Fusion **55** 115010.
[15] Murakami A et al 2012 Plasma Phys. Control. Fusion **54** 055006.
[16] Hosoi K et al 2013 Fusion Sci. Technol. **63** 244.
[17] Islam M M et al 2016 Plasma Fusion Res. **11** 2402126.
[18] Islam M M et al 2016 AIP Conf. Proc.**1771** 030018.
[19] Inutake M et al 1985 Phys. Rev. Lett. **55** 939.
[20] Nakashima Y et al 2011 J. Nucl. Mater. **415** S996.
[21] Yatsu K et al 1999 Nucl. Fusion **39** 1707.

The preparation and analysis of ceramic fibers with deposited catalytic particles by plasma technology

M Milieška, R Kėželis and V Snapkauskienė

Lithuanian energy institute, Plasma processing laboratory, Breslaujos str. 3, LT 44403 Kaunas, Lithuania

mindaugas.milieska@lei.lt

Abstract. In this work, the ceramic micro- and nanofibers were formed using atmospheric pressure thermal plasma reactor. The catalytic (copper, titanium) particles and their mixtures were deposited on the surface of zeolite fibers during their formation. The prepared catalytic fibers were analyzed using scanning electron microscopy (SEM), X-ray diffraction analysis (XRD), BET (Brunauer–Emmett–Teller) surface area measurements and Energy-dispersive X-ray spectroscopy (EDS). The SEM analysis showed that the obtained fiber diameters are in the range from 10 μm to few hundred nanometers. Some of the catalytic particles were deposited without changes and some were melted or burst. The XRD analysis showed that the zeolite fiber without the catalytic particles is amorphous as there are no crystal phase peaks. After the deposition of catalytic particles the peaks of the catalytic metals and their oxides appear. The BET method showed that the surface area of the catalytic fiber is 13-18 m2/g. The EDS analysis showed that the quantity of catalytic particles in the catalytic fiber is 7-18 %.

1. Introduction

The Carbon monoxide (CO), nitrogen oxides (NOX) and hydrocarbons (HC) are commonly targeted pollutants worldwide that have caused global warming and environmental pollution [1]. Efficient purification of such exhaust gases is generally carried out using noble metal catalysts such as platinum, rhodium and palladium. However, these noble metals are limited resources and reduction of their usage has become important for wider applications [2].

Zeolite catalysts have been widely used in various industrial field [1]. Usually they are used as supports for platinum as they have a porous structure that can confine molecules in small spaces. The zeolite affords large surface area and large cage with small window structure. This provides the space confinement and strong metal-support interaction for metal nanoparticles with high dispersion, good stability, and adjustable acidic strength when it is used as a catalyst support generally in the form of beads or pellets [3].

Fibrous catalysts made of inexpensive metals, carbons or glasses allow a larger field of applications instead of gauze catalysts woven of wires of fine precious metals (Pt, Ru, Ag). Fibers offer flexibility and endless forms, which do not apply to the usual powdered or granular materials. It is thus possible to adjust the geometric appearance of the fibrous substrates in order to fit it into a flue gas channel of arbitrary shape. Low pressure drop of processed gases, high surface area, safer operation and etc. are identified as other advantages of fiber catalysts [4].

The recently developed ceramic fiber production process using plasma-chemical reactors provides an opportunity to form fine fibers from any kind of ceramic materials by employing the kinetic energy of plasma flow [5]. Such microfiber (in some cases even up to 300 nm in thickness) can be used as catalyst supports for cheaper catalytically active metals like Cu, TiO_2 and etc. as its melting temperature is higher than the operating temperatures of the catalysts. Copper is usually used in the catalyst production as additive to Pd for CO oxidation [6], in selective catalytic reduction of NO_x and NH_3 [7] and etc. The titanium oxide is used in currently commercial catalysts based on V_2O_5–WO_3–TiO_2 [8] or in composites with platinum-gold metals for CO oxidation and NO_x reduction [9]. Chen et al. [10] showed that CuO/TiO_2 catalyst had superior catalytic activity with NO conversion and N_2 selectivity.

In this work, the ceramic fibers were formed using atmospheric pressure thermal plasma reactor. The catalytic (copper, titanium) particles were deposited on the surface of zeolite fibers during their formation. The

2. Methodology

The zeolite fiber used in this work was prepared using special plasma-chemical reactor, which in detail discussing the thermodynamic processes taking place in the reactor has been described in our other work [11]. The fiber was formed from the zeolite powder (up to 60 μm size).

As the obtained zeolite fiber in the catalyst is used as the support, the catalytically active particles (Cu, TiO_2) were deposited on the zeolite fiber during the fiber formation process. The deposition of catalytic particles was designed at the exhaust of the reactor where the formation of the fiber from the melt takes place spraying the catalytic particles on the surface of the formed fiber. Consequently, the catalytically active particles are deposited on the surface of the filaments of formed fiber and can react with the processed gases during catalytic process. The catalytic fiber production process in channel of the plasma chemical reactor is presented in figure 1 (where 1 – the feeder of ceramic particles; 2 – the feeder of catalytic particles; 3 – reactor walls cooled with water; 4 – ceramic melt flow; 5 – produced fibers and granules; G_1 – air plasma flow from plasma torch; G_2 – air flow for injection of ceramic materials; G_3 – air flow for spraying of catalytic particles; G_c – flow of catalytic particles; G_g – propane gas flow; G_z – flow of ceramic (zeolite) particles).

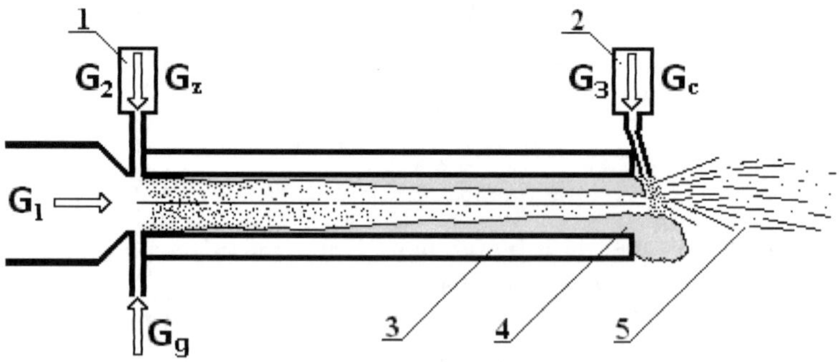

Figure 1. The schematic view of the catalytic fibre production process.

The channel of plasma-chemical reactor designed for ceramic fiber production was applied for formation of catalytic fiber by, as mentioned above, attaching the feeder of catalytic particles at the outlet nozzle of the reactor, rejecting the additional air flow rate which is usually used to increase the kinetic energy of plasma flow, decreasing the diameter of outlet nozzle of the reactor from 16mm to 13mm. The parameters of plasma chemical reactor are presented in Table 1.

The copper (Cu), titanium oxide (TiO_2) and their mixture were used as catalytically active particles. The mixture was prepared by mixing the equal weights of Cu and TiO_2 particles. The size of those particles was between 40-60 μm.

The prepared catalytic fibers were analyzed using scanning electron microscopy (SEM), X-ray diffraction analysis (XRD), surface area measurements and Energy-dispersive spectroscopy (EDS).

Table 1. The plasma-chemical reactor parameters.

Parameter	Value
Plasma torch power, kW	80
Air flow rate (G_1), 10^{-3} kg/s	22
Propane gas flow rate (G_g), 10^{-3} kg/s	1.2
Dispersive particle flow rate (G_z), 10^{-3} kg/s	1
Plasma flow temperature, leaving the reactor, K	3100
Plasma flow velocity, leaving the reactor, m/s	~1200
Plasma torch efficiency, %	80

3. Results and discussions

The morphology of formed ceramic fibre with and without catalytic particles is shown in Fig. 2. It can be seen, that the surface of formed fiber is smooth with some small derivatives on surface. The fiber diameters between 10-20 μm dominate. The surface area of such fiber is up to 6 m^2/g.

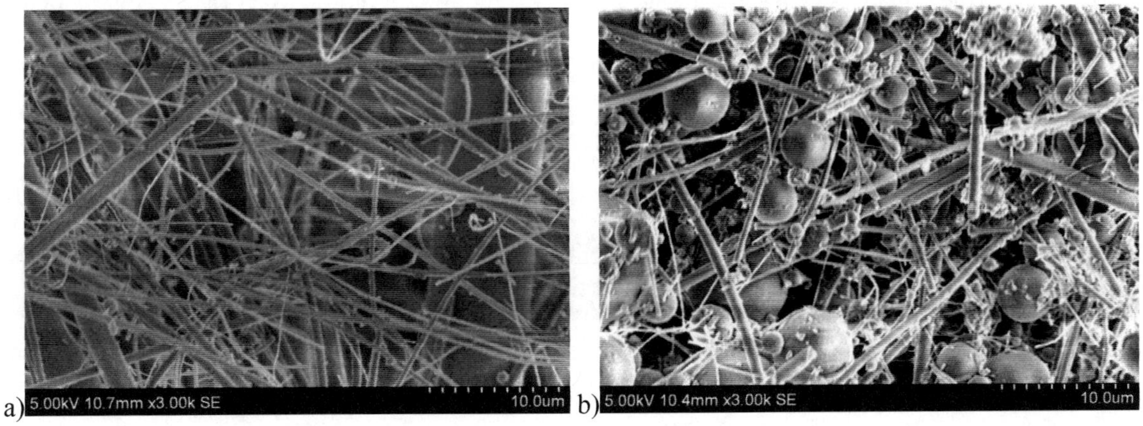

Figure 2. SEM images of obtained fiber. a) – without, b) – with catalytic particles (Cu).

The sprayed catalytic particles stick to the ceramic fiber at the end of the plasma-chemical reactor where plasma flow temperature is about 3100 K. However, the kinetic energy of plasma flow is very high, so the room temperature catalytic particles are not able to heat up to their melting point and are blown away with the ceramic fiber. In figure 2b it can be seen that the formed catalytic fiber consists of small (up to ~40 μm) size particles and ceramic fiber filaments which are the main component of pure zeolite fiber seen in figure 2a. Some of the catalytic particles were deposited without changes and some were melted or burst. The obtained fiber diameters are in the range from 10 μm to few hundred nanometers. The surface area of ceramic fiber with doped catalytic particles increases from 6 m^2/g in case of zeolite up to 18 m^2/g after addition of the TiO_2. The addition of Cu increases the surface area up to 13 m^2/g.

It was found that zeolite fiber prepared using plasma technology has almost only amorphous phase mullite there are no crystalline phase peaks in the XRD patterns. The XRD patterns of the fiber samples with doped Cu particles (figure 3a) show the crystalline form of Cu, corundum (crystalline form of Al_2O_3), quartz (crystalline form of SiO_2) and synthetic mullite ($Al_{4.64}\,Si_{1.36}O_{9.68}$). The sample with doped TiO_2 particles (figure 3b) consists of corundum, anatase, and rutile, the crystalline form of quartz and mullite ($Al_{2.3}Si_7O_{4.85}$) and amorphous mullite. The addition of catalytic particles has the influence on the mullite as part of it experiences conversion in phase from amorphous to crystalline. As the crystalline variety shrinks less than the amorphous variety it is an advantage for such catalyst supports [3].

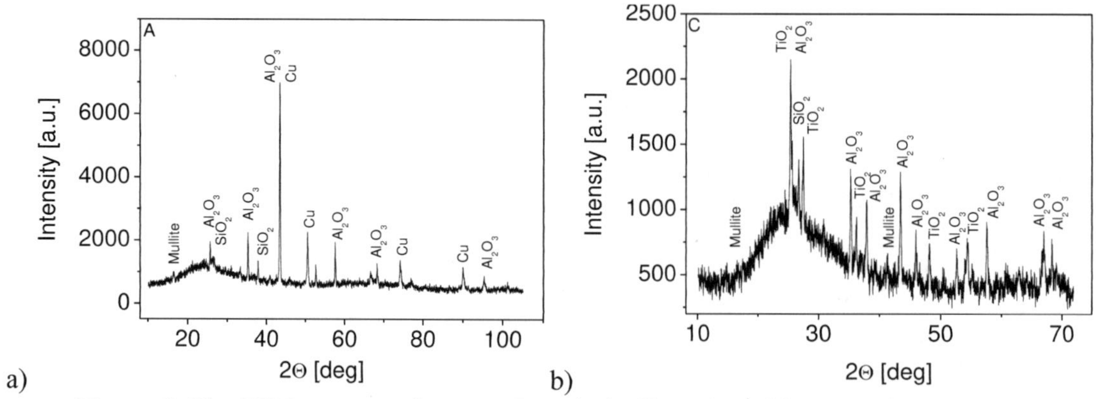

Figure 3. The XRD patterns of prepared catalytic fiber. a) – with Cu particles; b) – with TiO_2 particles

The EDS analysis showed that the quantity of catalytic particles in the catalytic fiber in case of Cu is 6.7 %, while the dominating species are those which compose the zeolite: oxygen – 53.6 %, silicon – 24.1 and aluminum – 12.3 % in normalized atomic concentration. The impurities amount for 3.3 %. In case of TiO_2 particle injection, there are 18.1 % of catalytic particles in the fiber. The oxygen amount in the fiber is 51.3 %, silicon – 21.4 %, aluminum – 4.8 % and the impurities – 4.4 % of normalized atomic concentration. The fiber without catalytic particles consist of 55 % of oxygen, 26 % of silicon, 14.6 % of aluminum and 4.4 % of impurities.

Conclusions

The catalytic ceramic fiber can be obtained by spraying the catalytic particles on the fiber during its formation. The addition of catalytic particles has the influence on zeolite fiber as part of it converts from amorphous to crystalline phase. In case of TiO_2 and Cu doping, the surface area of the fiber increases up to 3 and 2 times and the catalytically active media amounts for 18.1 % and 6.7 % in the fibrous catalyst, respectively. The further direction of the research – catalytic efficiency of prepared fiber.

References

[1] Wu S H, Hsieh C C, Chiang C C, Horng J J, Pan W P and Shu C M 2012 *J. Therm. Anal. Calorim.* **109** 945
[2] Koga H, Umemura Y, Ishihara H, Kitaoka T, Tomoda A, Suzuki R and Wariishi H 2009 *Appl. Catal., B* **90** 699
[3] Chu Y, Zhang B, Zhang Q, Wang Y, Su D S and Wei F 2013 *Microporous Mesoporous Mater.* **169** 201
[4] Matatov-Meytal Y and Sheintuch M 2002 *Appl. Catal., A* **231** 1
[5] Kėželis R, Valinčius V, Grigaitienė V, Milieška M and Mėčius V 2010 *Proc. Int. Conf. on Gas Discharges and Their Applications (Greifswald)* vol 1 (Friedland:Germany/Steffen GmbH) p 360
[6] Wang F and Lu G 2010 *Int. J. Hydrogen Energy* **35** 7253
[7] Pereda-Ayoa B, De La Torrea U, Illán-Gómezb M J, Bueno-Lópezb A and González-Velasco J R 2014 *Appl. Catal., B* **147** 420
[8] Chen L, Li J and Ge M 2011 *Chem. Eng. J.* **170** 531
[9] Song G H, Shi M Q, Chu Y Q and Ma C A 2013 *Electrochim. Acta* **112** 53
[10] Chen H, Zou H, Lu Z. and Wu J 2012 *Adv. Mat. Res.* **550-553** 225
[11] Milieška M, Kėželis R, Valinčius V and Grigaitienė V 2013 *Proc. Int. conf. on Experimental Heat transfer, Fluid Mechanics and Thermodynamics (Lisbon)* vol 1, (Lisbon:Portugal/Edicao Electronica Lda.), p 1

IOP Conf. Series: Journal of Physics: Conf. Series **959** (2018) 012009 doi:10.1088/1742-6596/959/1/012009

Molecular ND Band Spectroscopy in the Divertor Region of Nitrogen Seeded JET Discharges

E Pawelec[1], T Dittmar[2], A Drenik[3], A Meigs[4] and JET contributors[1]

EUROfusion Consortium, JET, Culham Science Centre, Abingdon, OX14 3DB, UK
[1]University of Opole, Institute of Physics, Oleska 48, Opole, Poland
[2]Forschungszentrum Jülich GmbH, Institut fr Energie- und Klimaforschung – Plasmaphysik, Partner of the Trilateral Euregio Cluster (TEC), 52425 Jülich, Germany
[3]Max-Planck-Institut fr Plasmaphysik, 85748 Garching b. Mnchen, Germany
[4]UKAEA-EURATOM Association, Culham Science Centre, Abingdon, OX14 3DB, UK

E-mail: ewap@uni.opole.pl

Abstract. In this contribution we present OES measurements in the JET tokamak of the deuterated NH (ND) radical and the correlation between results of those experiments and measurement of ammonia production. The observation region covers most of the divertor and its outer throat. Measurements are performed in different magnetic configurations. The results include temporal and spatial dependence of the molecular emission intensity and study of the emission band shape (vibrational and rotational temperatures) during different JET pulses, with or without nitrogen seeding. Results are a step towards the understanding of nitrogen-containing molecule creation and destruction in the divertor plasma.

1. Introduction

Studying the nitrogen plasma impurity is important in fusion plasmas, as this gas is both used for divertor temperature control and has an apparent beneficial effect on confinement. However, it reacts with hydrogen, forming molecules, also stable ones, of which the most prominent is ammonia. In DT plasmas, formation of tritiated ammonia poses several potential issues, e.g. the retention of tritium-containing molecules on metallic surfaces and in the divertor cryo-pumps.

Optical emission spectroscopy (OES) of ammonia is very difficult, because this molecule does not have any strong bands in the visible region. Nevertheless, if by OES we can determine where, when and in which quantity relevant hydride molecules (radicals) are created, the results may give an estimation of the ammonia creation, as those radicals are both precursors for creation of ammonia and results of its destruction. The strongest of hydride bands is a NH radical (or its isotope equivalent) $A^3\Pi$—$X^3\Sigma-$ $\Delta v = 0$ near-UV band at ≈335 nm. Each vibrational transition of this band consists of line-like Q rotational branch and much more distributed P and R branches, with band head at short wavelength. This transition was observed in JET during first experiments with nitrogen seeding in the ITER-Like Wall (ILW) configuration [1].

[1] See the author list of X. Litaudon et al 2017 *Nucl. Fusion* **57** 102001

IOP Conf. Series: Journal of Physics: Conf. Series **959** (2018) 012009 doi:10.1088/1742-6596/959/1/012009

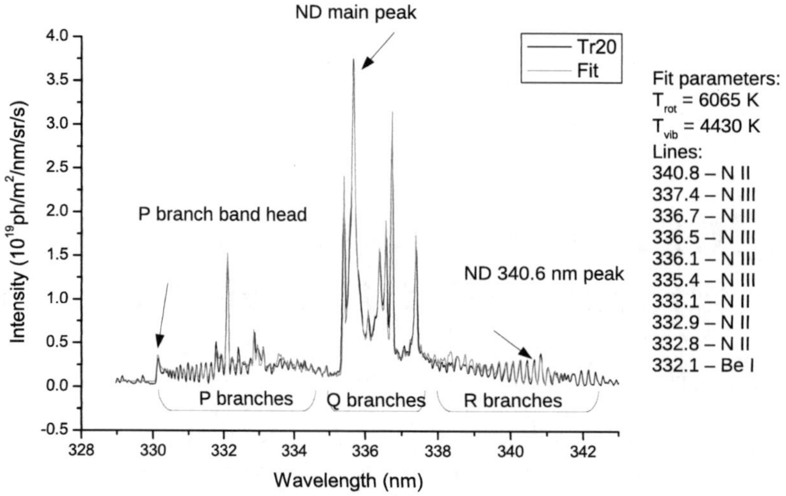

Figure 1. Experimental spectrum (JPL #92143, R = 2.88 m, t = 12.34 s) and fit (most important parameters are shown on the figure).

2. Experiment

Spectra were registered using mirror-link spectrometer KT3 (described e.g. in [2]). This spectrometer has 22 lines of sight covering divertor and part of the outer throat. In the nitrogen seeding experiments the UV detector and 1200 l/mm grating were used, to detect at least partially resolved rotational structure of the ND molecular band.

Experiments were conducted with two different magnetic field configurations: horizontal target, where the outer strike point is on the divertor plate and vertical target, where the strike points are both on divertor throat plates (described in e.g. [3]). In both configurations the ND spectra were visible, although with different intensities. Nitrogen seeding was starting after the formation of the X-point and switching on the NBI heating. Some nitrogen (and ND) was visible before seeding started, and this is called in this contribution a "legacy" phase, where the nitrogen present in the plasma is a "legacy" from previous pulse.

3. Synthetic spectrum

Synthetic molecular spectrum was created with the help of PGOPHER code [4] with diatomic molecular data taken from [5]. Vibrational transitions from 0-0 to 3-3 were taken into account. The exported wavelengths, energy levels and Einstein factors for each transition were used for creating a fitting procedure for the experimental spectrum. Free parameters included: rotational and vibrational temperatures, total molecular band intensity, background level and intensities of atomic/ion lines overlapping the molecular band. An example of the fit is shown in figure 2. Including 10 nitrogen and beryllium lines is sufficient for such strong ND spectrum, but the number of lines has to be increased if recordings with relatively weaker ND bands are to be fitted – there is also a weak N III line overlapping the main 0-0 Q branch.

Fortunately, fitting the whole spectrum may be not necessary for estimating the band intensity or temperatures. Study of the experimental and synthetic spectrum reveal some features which can be used for quick estimation of those values. In the case of ND transition obvious features for total band intensity are amplitudes of Q branches of the vibrational transitions (especially 0-0), but as figure 2 shows, they depend on both rotational and vibrational temperatures. Also, there are many nitrogen lines overlapping the center peaks (figure 1). A better choice is one of composite peaks from R branch, located at 340.6 nm, which amplitude for temperatures above

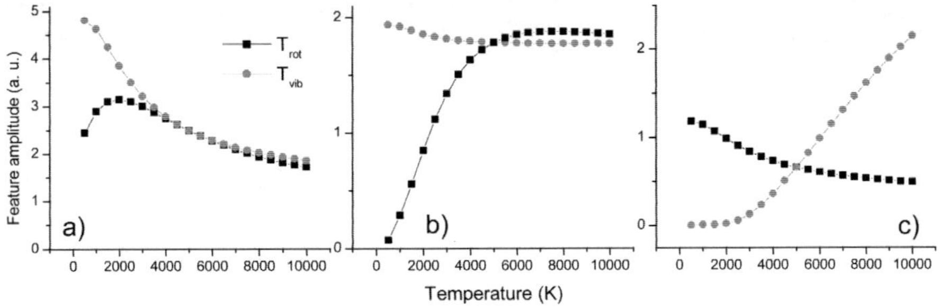

Figure 2. Temperature dependence of a a) main Q branch peak amplitude, b) 340.6 nm feature c) P band head to 340.6 nm feature ratio. For each temperature curve the other temperature is set at 5000 K.

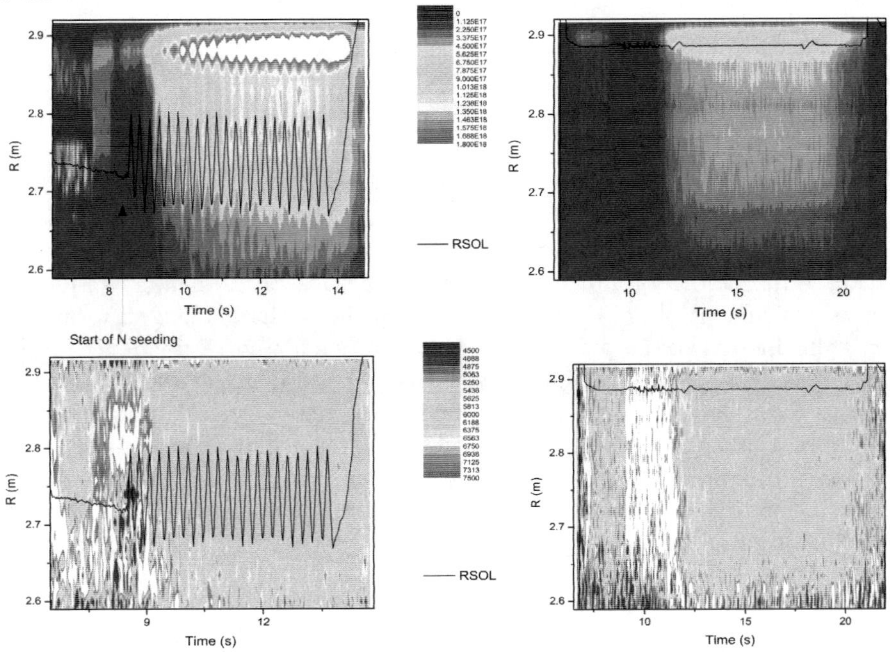

Figure 3. Top: intensity distribution for HT (left) and VT (right) configurations. Bottom: corresponding rotational temperatures for both experiments. Black line (RSOL) denotes position of the outer strike point.

4000 K is nearly independent on any of the temperatures. Short wavelength band head (P branch of 0-0) depends very strongly on rotational temperature, and its ratio to the 340.6 nm peak can be used for T_{rot} estimation.

4. Results

Estimations of total band emission (which is proportional to density of excited molecules) and rotational temperatures were made from amplitudes of the chosen spectral features (340.6 nm and 330.1 nm) with the background estimated from radiation at 329.9 nm. Resulting spatio-temporal distributions for JPL #92143 and #92503 are shown in figure 3. The density of excited ND radicals is much higher in the case of horizontal target experiment, the band intensity is higher by factor of 3. In both cases the intensity is highest at the outer throat of the divertor,

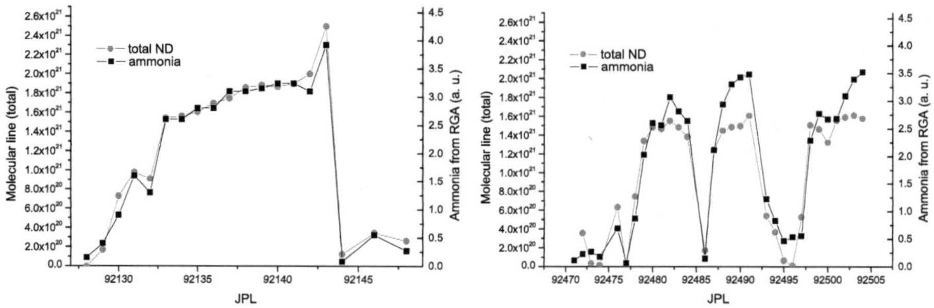

Figure 4. Total ND intensity during JET pulse compared to total ammonia production during the pulse. Left – horizontal target experiment, right – vertical target experiment.

where the N_2 is puffed from. There is a dip in emission between the divertor plates and the throat, and definite dependence on the location of the strike point or, in fact, the separatrix, as the recorded band intensity is a result of the integration along the spectrometer line of sight. The intensity is highest in the seeded part of the pulse, but there is also a time during and just after the X-point formation, when the strike point glides along the vessel walls releasing the adsorbed nitrogen, when ND band emission is visible as well (legacy phase). The rotational temperature distribution is remarkably uniform – for most places and times T_{rot} is between 5600 and 6300 K. There seems to be a lower temperature region just outside the separatrix visible in the horizontal target experiments, but the explanation of this phenomenon is still unclear.

Main value of ND radical plasma content measurement is that it can serve as an indicator for the ammonia production in the plasma. To show that such a correspondence exists the total (spatially and temporally integrated) ND intensity was compared to the total ammonia registered by RGA (Residual Gas Analyzer, [5]), keeping the same scale for vertical and horizontal target experiments. Figure 4 shows, that the results are promising – the ratio is nearly constant and only for a few pulses difference exceeds 10 %. Those discrepancies can be explained taking into account, that the observed region is only a part of area where molecules can be created and in the case of vertical target relatively more molecules can be created in unobserved region.

5. Conclusions

Observation and analysis of nitrogen seeded pulses in JET show that the ND radical radiation may be used as an estimate of ammonia production. Analysis of the synthetic spectra yields also a procedure for a quick estimation of T_{rot} and total intensity of ND band, at least in conditions of JET divertor region.

Acknowledgements

This work has been carried out within the framework of the EUROfusion Consortium and has received funding from the Euratom research and training programme 2014-2018 under grant agreement No 633053. The views and opinions expressed herein do not necessarily reflect those of the European Commission. This work is also supported by Polish Ministry of Science and Higher Education for 2017 as co-financed international project.

References

[1] Oberkofler M et al 2015 *Fusion Eng. Des.* **98–99** 1371–74
[2] Meigs A G et al 2013 *J. Nuc. Mat.* **438** S607-11
[3] Tamain P et al 2015 *J. Nuc. Mat.* **463** 450-4
[4] Western C M 2016 *J. Quant. Spectrosc. Radiat. Transfer* **186** 221–42
[5] Patel-Misra D, Sauder D G, and Dagdigian P J 1990 *Chem. Phys. Lett.* **174** 113–8
[6] Drenik A et al. 2017 *Fusion Eng. Des.* [Internet] May 31.

Proton acceleration in two-species targets irradiated by an ultra-intense femtosecond laser pulse

J Domański[1], J Badziak and S Jabłoński

Institute of Plasma Physics and Laser Microfusion,01-497 Warsaw, Poland

[1] jaroslaw.domanski@ifpilm.pl

Abstract. This paper presents results of two-dimensional particle-in-cell simulations of proton acceleration at the interactions of a 130 fs, linearly polarized laser pulse of intensity from the range 10^{21} –10^{23} W/cm^2, predicted for the ELI (Extreme Light Infrastructure) lasers, with thin hydrocarbon (CH) or hydride (ErH_3) targets. It is shown that for the targets of the areal density σ > 0.1 mg/cm^2 and laser intensities below 10^{22} W/cm^2 a higher efficiency of proton acceleration is achieved for hydride targets. However for the highest, ultra-relativistic laser intensities (~ 10^{23} W/cm^2) considerably higher proton energies and proton beam intensities are achieved for thin ($\sigma \leq 0.1$ mg/cm^2) CH targets. In this case, at short distances from the irradiated CH target (< 50 um), the generated proton pulses are very short (< 20 fs), and the proton beam intensities and the proton current densities reach extremely high values, > 10^{21} W/cm^2 and > 10^{12} A/cm^2, respectively, which are much higher than those attainable in conventional accelerators. Such proton beams can open the door for new areas of research in nuclear physics and high energy-density physics as well as can be useful for materials research.

1. Introduction

Extreme Light Infrastructure (ELI) is a currently implemented European project that uses cutting-edge laser technologies to build multi-PW laser systems generating femtosecond pulses of ultra-relativistic intensities (10^{22}-10^{23} W/cm^2).[1] The laser pulses of such parameters will have a potential to produce intense ion beams of sub-GeV and GeV energies, demanded for applications in various branches of nuclear physics[2] and high energy-density physics[3-5]. However studies of laser-driven ion acceleration at such high laser intensities are in very initial stage [6-9] and need to be intensely developed. The ions can be accelerated by several laser-induced mechanisms including the target normal sheath acceleration (TNSA)[4,10] or the radiation pressure acceleration (RPA)[4,6,10,11]. Which of these mechanisms are dominant depends on the laser-target interaction condition. For this reason, the target and the laser pulse parameters should be carefully selected to produce the ion beam of characteristics required for a particular application.

This paper presents results of two-dimensional (2D3V) particle-in-cell simulations of proton acceleration at the interaction of a 130 fs, linearly polarized laser pulse of intensity from the range 10^{21}-10^{23} W/cm^2 and 800 nm wavelength, predicted for the ELI lasers, with thin hydrocarbon (CH) or hydride (ErH_3) targets (the areal targets density ranging from $\sigma_t = 0.06\ mg/cm^2$ to $\sigma_t = 0.8\ mg/cm^2$). The effect of the laser pulse intensity and the target thickness on the proton energy spectrum, the proton beam spatial distribution and the proton beam shape and intensity is investigated.

2. Results and discusion

Figure 1 presents a two-dimensional spatial distribution of the charge density of protons (a,d), heavy ions (b,e) and electrons (c,f) generated from both types of investigated targets. The targets were irradiated by the linearly polarized laser pulse of intensity equal to $2*10^{23}$ W/cm^2 and the presented situation correspond to the final stage of ion acceleration (the simulation time is equal to 0 0.4 ps). It can be seen that protons are clearly separated from the erbium ions which attain much lower velocities and most of these ions are concentrated close to the initial target position. As opposed to this case, the difference between velocities of protons and carbon ions is much smaller and some carbon ions move with velocities close to the proton velocities. For this reason for the CH target electrons are concentrated mainly in the region occupied by both proton and carbon ions. This effect leads to the longer opacity of the target and domination of the RPA mechanism than for the ErH_3 target. In effect, for the CH target the proton beam of higher mean energy and lower energy deviation is produced.

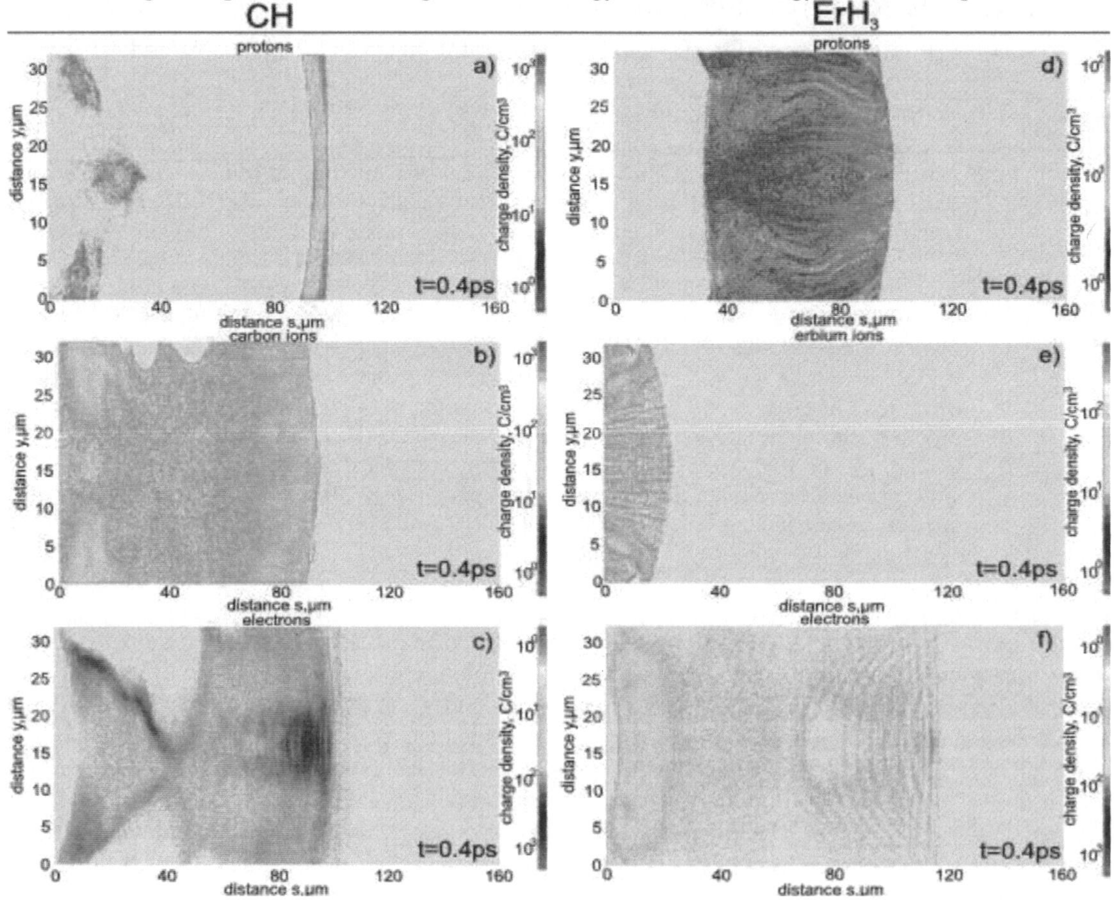

Figure 1. . 2D spatial distributions of charge density of protons (a, d), heavy ions (b,e) and electrons (c, f) for CH (a,b,c) and ErH_3 (d,e,f) targets. The targets were irradiated by the linearly polarized laser pulse of intensity equal to $2*10^{23}$ W/cm^2. t=0.4ps, σ_t =0.06mg/cm^2.

The mean energy of protons for both types of investigated targets as a function of laser beam intensity (a) and areal target density σ_t (b) is shown in figure 2. It can be observed that the influence of the target's compositions on the efficiency of ion acceleration significantly depends on the laser beam intensity and the target thickness. For high laser intensity $\geq 10^{22}$ W/cm^2 and thin targets of $\sigma_t < 0.1$ mg/cm^2, the mean proton energies are higher for the CH than for the ErH_3, while for the lower intensities or thicker targets the higher efficiencies of proton acceleration are achieve for the erbium hydride targets. This effect could be explained by the fact that at lower laser intensities and thicker targets the influence of relativistic transparency on the acceleration process is low (the ErH_3 targets become transparent for light later) and the efficiency of proton acceleration is determined by the q/m value of the heavy ion component of the target [12,13]. Furthermore, it is worth mentioning that for the highest intensity of the laser and thinnest polystyrene target the proton beam of mean energy of 1.8GeV and the energy dispersion $dE/\bar{E} = 0.3$ is produced.

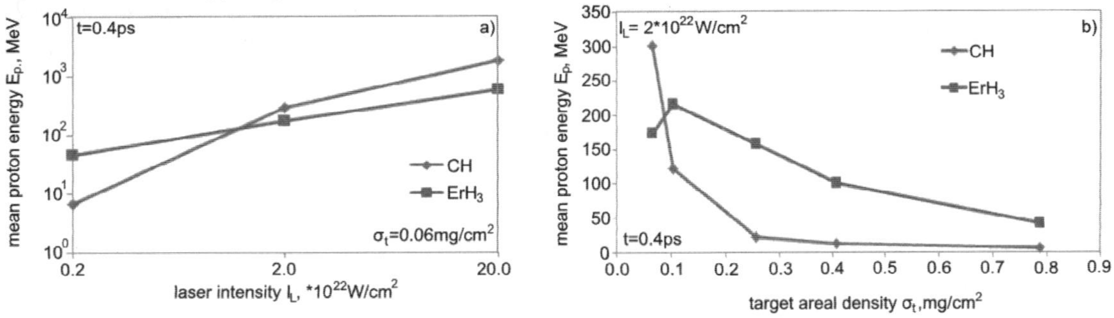

Figure 2. The quantitative values of the mean energy for protons as a function of the laser beam intensity and the areal targets density. t=0.4ps.

The characteristics of ion beams such as the beam intensity, shape and duration of the ion pulse are important from the point of view of various potential applications of laser-accelerated ions [2,4,10]. Figure 3 presents the temporal distribution of intensity of the proton beam generated from the polystyrene and erbium hydride targets (a) and the peak proton beam intensity as a function of laser beam intensity (b). These intensities were recorded 40 μm behind the rear surface of the target. The target's composition has a significant influence on this characteristic. In the case of higher laser intensities the peak proton beam intensities are higher for the CH target, while in the case of lower laser intensities they are higher for the ErH_3 target. In particular, at laser pulse intensities of $2*10^{23}$ W/cm^2 and the CH target the proton beam of duration about 11fs and peak intensity of $8*10^{21}$ W/cm^2 is produced.

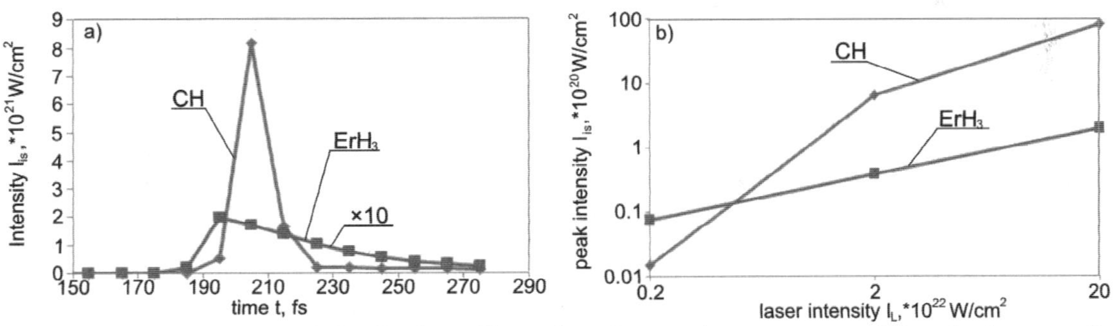

Figure 3. The temporal distribution of intensity of proton beams generated from both types of investigated targets by the laser pulse of intensity $2*10^{23}$ W/cm^2 (a) and the peak proton beam intensity as a function of laser beam intensity (b).

Figure 4 presents the temporal run of the total current (integrated over the whole transverse size of the simulation box) at the distance of 40 μm from the rear surface of the target. It can be seen that especially at the time longer (at least several times) than the laser pulse duration the total current is dominated by electrons (the current is negative). However, the strong influence of the target's composition on the distribution of the total current is visible. For the ErH_3 target, the value of current is equal to 1MA or less, while for the polystyrene target it reaches the value of several MA. This strong flux of fast electrons escaping from the target is the cause of the creation of a net positive charge of the target. The neutralization current flowing through the target and the target holder can be a source of a strong electromagnetic pulse (EMP), usually in the GHz domain, which propagates into the interaction chamber and may be harmful for diagnostic equipment and other electronic devices [14,15]. The problem of EMP generated in the laser-target interaction is especially important for big laser facilities like ELI and is currently an intensely studied target. Our simulations, in particular the results presented in figures 4, show briefly what way the fundamental source of EMP is created in the sub-ps time scale when the ion beam is formed and accelerated due to the laser-target interaction. More detailed analysis of this issue is beyond the scope of the present paper and will be done in other our papers.

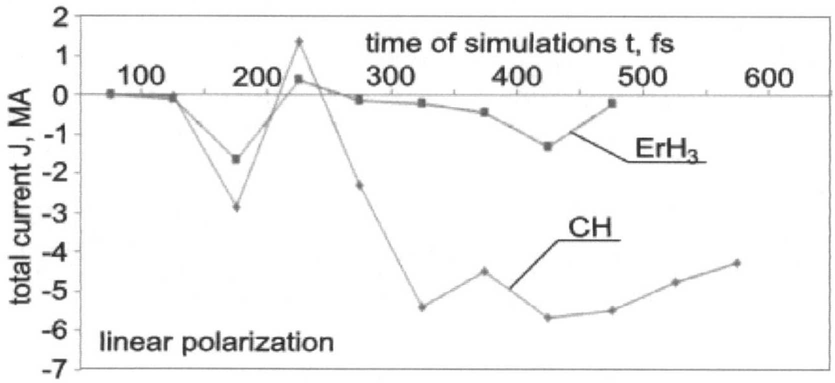

Figure 4. The temporal run of net total current recorded at the distance of 40 μm from the target rear surface. σ_t =0.06mg/cm^2, $I_L = 2*10^{23}$ W/cm^2.

3. Conclusion

In conclusion, it is shown that for the targets of the areal density >0.1 mg/cm^2 and laser intensities below 10^{22} W/cm^2 a higher efficiency of proton acceleration is achieved for the hydride targets. However, for the highest, ultra-relativistic laser intensities ($\sim 10^{23}$ W/cm^2) considerably higher proton energies and proton beam intensities are achieved for thin ($\leq$0.1 mg/cm^2) CH targets. In this case, at short distance from the irradiated polystyrene target (<50 μm), the generated proton pulse is very short (<20fs) and proton beam intensity reaches extremely high values >10^{21} W/cm^2, which is much higher than those achievable in conventional RF-driven accelerators. Such proton beams can open the door for new areas of research in nuclear physics and high energy-density physics as well as prove useful for applications in materials science.

References

[1] Danson C, Hillier D, Hopps N and Neely D 2015 *High Power Laser Sci. Eng.* **3** e3
[2] Ledingham K W D and Galster W 2010 *New J. Phys.* **12** 045005
[3] Patel P K, MacKinnon A J, Key M H, et al. 2003 *Phys. Rev. Lett.* **91** 125004
[4] Macchi A, Borghesi M, and Passoni M 2013 *Rev. Mod. Phys.* **85** 751
[5] Fernandez J C, Albright B J, Beg F N, et al. 2014 *Nucl. Fusion* **54** 054006
[6] Esirkepov T, Borghesi M, Bulanov S V, et al. 2004 *Phys. Rev. Lett.* **92** 175003
[7] Yin L, Albright B J, Hegelich B M, et al. 2007 *Phys. Plasmas* **14** 056706
[8] Zheng F L, Wang H Y, Yan X Q, et al. 2012 *Phys. Plasmas* **19** 023111
[9] Xu Y, Wang J, Qi X, et al. 2017 *Phys. Plasmas* **24** 033108
[10] Daido H, Nishiuchi M and Pirozhkov A S 2012 *Rev. Prog. Phys.* **75** 056401
[11] Badziak J, Jabłoński S, Parys P, et al. 2008 *J. Appl. Phys.* **104**, 063310
[12] Domański J, Badziak J and Jabłoński S 2016 *Laser Part. Beams* **34** 294-98
[13] Domański J, Badziak J and Jabłoński S 2017 *Laser Part. Beams* **35** 286-93
[14] Dubois J L, Lubrano-Lavaderci F, Raffestin D, et al. 2014 *Phys. Rev. E* **89** 013102
[15] Poye A, Dubois J L, Lubrano-Lavaderci F, et al. 2015 *Phys. Rev. E* **92** 043107

Fixation of the stressed state of glass plates by coating them with thin films using a plasma focus installation

V.N. Kolokoltsev[1], V.F. Degtiarev[2], I.V. Borovitskaya[1], V.Ya. Nikulin[3,4], E.N. Peregudova[3], P.V. Silin[3], and A.A. Eriskin[3]

[1] A. A. Baikov Institute of Metallurgy and Material Science of RAS, Leninsky pr. 53, 119991 Moscow, Russia

[2] Moscow Technical University of Communications and Informatics, Aviamotornaya ul. 8a, 111024 Moscow, Russia

[3] P. N. Lebedev Physical Institute of RAS, Leninsky pr. 53, 119991 Moscow, Russia

[4] National Research Nuclear University "MEPhI," Moscow, 115409 Russia

E-mail: vnik@sci.lebedev.ru; symp@imet.ac.ru

Abstract. Elastic deformation in transparent mediums is usually studied by the photoelasticity method. For opaque mediums the method of film coating and strain gauge method are used. After the external load was removed, the interference pattern corresponding to elastic deformation of the material disappears. It is found that the elastic deformation state of the thin glass plate under the action of concentrated load can be fixed during the deposition of a thin metal film. Deposition of thin copper films was carried out by passing of plasma through the copper tube installed inside the Plasma Focus installation. After removing of the load, interference pattern on the glass plates was observed in the form of Newton's rings and isogers in non-monochromatic light on the CCD scanners which uses fluorescent lamps with cold cathode. It is supposed that the copper film fixes the relief of the surface of the glass plate at the time of deformation and saves it when the load is removed. In the case of a concentrated load, this relief has the shape of a thin lens of large radius. For this reason, the interference of coherent light rays in a thin air gap between the glass of the scanners flatbed and the lens surface has the shape of Newton's rings. In this case, when scanning the back side of the plate, isogyres are observed. The presented method can be used in the analysis of the mechanical stress in a various optical elements.

1. Introduction

Determination of mechanical strain (stress) in solids is an urgent task. Elastic deformation in transparent media are usually investigated by photoelasticity [1-4]. Film coating techniques or moir fringes are used for non-transparent media [5-6]. Strain methods are laborious and provided information about the deformations in a limited number of points [7]. Methods of optical ellipsometry are complex and require a certain skill when analyzing the measurement results [8]. Measurement of deformations in samples is usually performed at the time of application of force. In the case of elastic deformation after unloading samples they return to its original state and the stress distribution pattern disappears. There are a limited number of methods to observe the deformation pattern in a solid after removal of the load. The photoelasticity method is based on samples of photosensitive transparent plastics, in which the heating and subsequent cooling leads to "freezing" of picture of internal stresses. [2,6]. By this method by cutting the samples into thin plates the stress distribution inside the samples is determined. In the method of moir fringes, on the contrary, the studied model is cut into thin plates, and then the optical grating is applied to each of the plates. Collected from the plate model is subjected to external forces, and then by studying the deformation of moir fringes that existed at the time of loading. Collected model from the plates is subjected to external forces, and then by studying the deformation of moir fringes that existed at the time of loading. The aim of this work is to study the pattern of the "frozen" (fixed) state deformation (stress) in the thin glass plates in situ by deposition of a thin metal film on a plasma focus installation.

2. Experimental methods

For the study, glass plates of thickness 1.5-3 mm and in size 40x40 mm were used. Before the experiment the glass samples were degreased in alcohol. Samples were subjected to elastic deformation by use of a special device (Figure 1), which was placed in a discharge chamber of plasma focus (PF-4, Lebedev Phys. Inst,). Applying force $\sim$ 1 Pa for all glass plates was the same. The plates were pressed against the surface of the sample holder at the corners by means of copper straps.

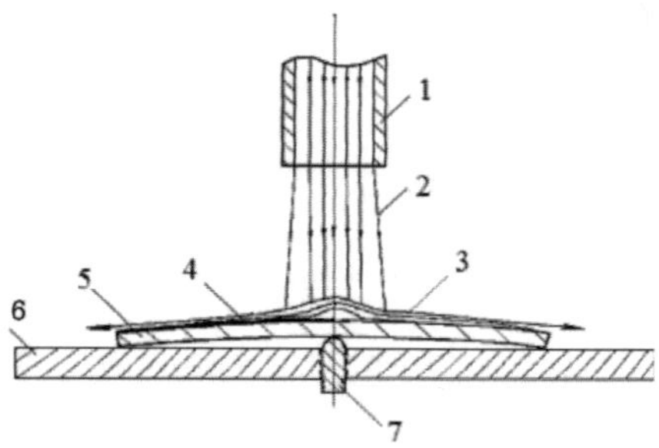

Figure 1. The device for monitoring "frozen" picture of deformation (stress) in a glass plate on the Plasma Focus Installation: 1 - copper tube; 2 - of plasma jet with the metal vapor; 3 - plasma region spreading over the surface of the glass plate; 4 - copper film; 5 - glass plate; 6 - sample holder (steel Art.20); 7 - the screw, creating a concentrated load to the glass plate.

The spraying of thin copper film was carried out under the scheme based on the plasma chemical reactor PF setting [9]. The discharge chamber was filled by argon gas at a pressure in the chamber $\sim$ 1 Torr. The stored energy in the capacitor bank was $\sim$ 3.6 kJ. The quarter of period of the electric discharge was equal $\sim$ 3 μs. The plasma jet was passed through the copper tube of 8.0 mm diameter and 70 mm in length. The distance between the edge of the copper tube and the surface of the glass plate was equal 28 - 30 mm. Received copper films were conductive and semi-transparent for the light. The electrical surface resistance of the films was $\leq 10^4 - 10^5$ Ohms/square. The structure of the films had the insular character. The observing the interference pattern on the surface of glass plates was carried out in non- monochromatic light on the CCD scanners with fluorescent lamps with a cold cathode [10]. Scanners of following models: HP Scanjet 3800, Epson Perfection 1650 Photo and Cano Scan D6604 were used to control observing the interference pattern. Due to the fact that the interference patterns were identical in the future we use only the scanner HP Scanjet 3800. The CIS scanners (Contact Image Sensor) on the LED strip were not considered in the work, because when they are used, the interference pattern is not turned.

IOP Conf. Series: Journal of Physics: Conf. Series **959** (2018) 012011 doi:10.1088/1742-6596/959/1/012011

3. Experimental results

Figure 2a shows a typical interference pattern on the original glass plate, which was scanned by the scanner HP Scanjet 3800. The character of the pattern indicates some deformation of the surface of the glass plates and the presence of the air gap in the form of a wedge between it and the plane of the document table glass, which leads to an interference pattern in the form of strips of equal thickness. Figure 2b shows a glass plate with a film deposited on the PF no-load applications. It is seen that the interference region around the spray pattern is not observed.

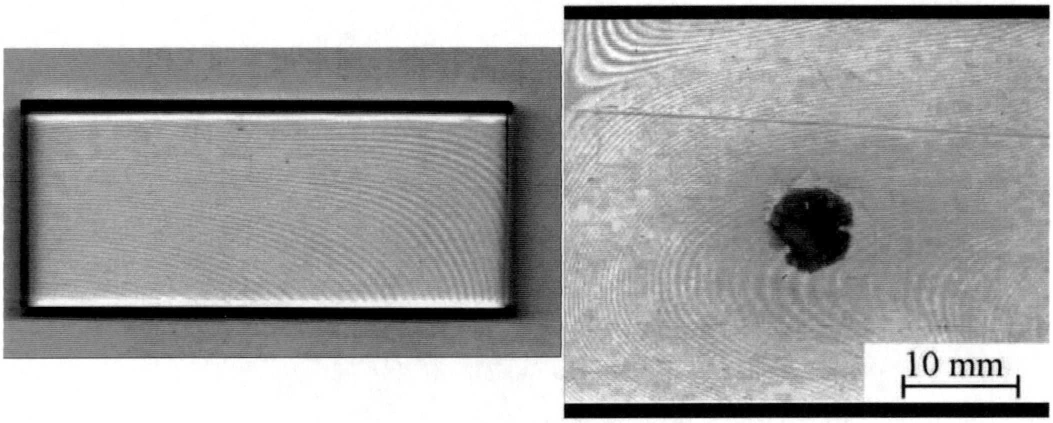

Figure 2. Interference pattern on the original glass plate (a) and on the plate exposed to the PF plasma pulse without the application of external force (b).

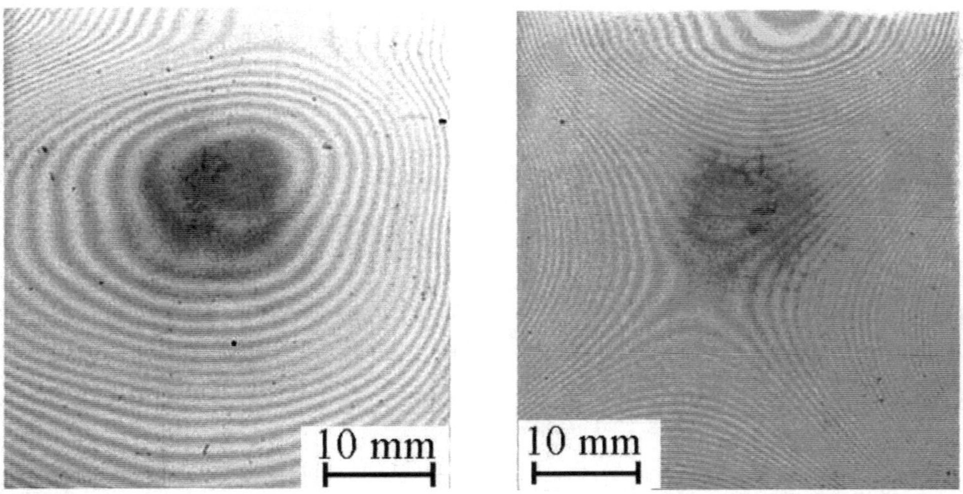

Figure 3. Interference picture on a glass plate, after removal of the load. a - on the side of a copper film - as Newton rings; b - on the back side (on the side of a concentrated force application) - as a hyperbolic bands.

Figure 3 shows the interference pattern on the glass plate after spraying copper film at the

IOP Conf. Series: Journal of Physics: Conf. Series **959** (2018) 012011 doi:10.1088/1742-6596/959/1/012011

time of application of a force. When scanning the plate on the sputtered film side the Newton interference rings are observed, and on the reverse side - interference fringes are seen in the form of hyperbolas disposed symmetrically relative to the force application point to the glass plate. The quality of the picture is significantly affected by the intensity of the plasma jet passing through the copper pipe. At high intensity of plasma the glass is destroyed and the observation of the interference pattern becomes difficult (Figure 4).

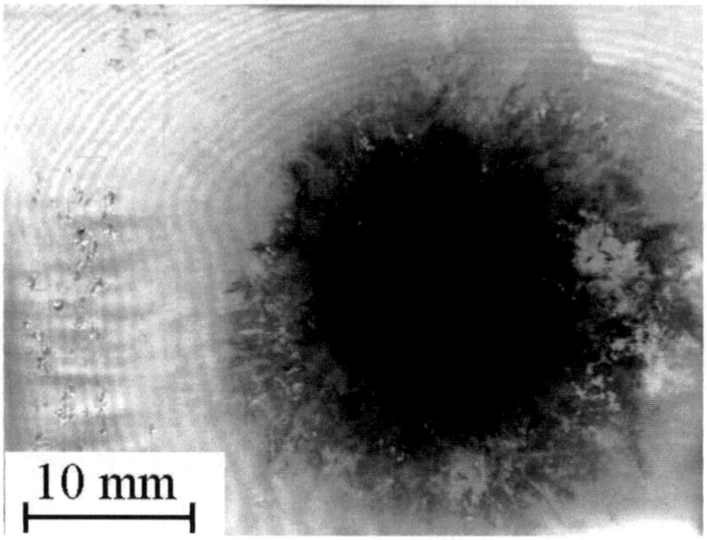

Figure 4. The interference pattern on a glass plate at a high intensity plasma jet.

4. Discussion of the results

The interference pattern in the form of Newton rings (Figure 3a) is typically observed in coherent interactions of light rays incident normal to the thin lens surface lying on the glass plate surface [11]. The role of the thin-lens in this case carries a metal film on the surface of the glass plate, curved in the form of a hemisphere. The solution of a particular problem for a semi-infinite thin plate under the action of a concentrated force is considered in [12]. It is known that the interaction of electromagnetic waves with the metal surface and the passing phase of the reflected waves are determined by the surface impedance and are 0 or $\pi/2$ [13]. However, in the case of "bad" conductors permittivity depends weakly on its complex parts, because the $\sigma \to 0$.

In this case, the thin copper film thickness of the wavelength order, and the character had islet surface resistance $\sim 10^4 - 10^5$ ohms / square. A feature of the observed interference pattern (Figure 3a) is the emergence of strong intensity interference rings in the green region of the spectrum and more faint rings pink. The maximum transmission of thin films of Cu is in the range $\lambda = 0{,}558\ \mu$ (green-yellow region of the spectrum), the near-infrared region of the spectrum is a strong reflection of the long-wave radiation. It can be assumed that the interference associated with the coherent interaction of light rays reflected from the copper film with rays of light passing through the film deep into the glass plate.

Currently, the nature of "freezing" the deformation structure in a thin glass plate during deposition of the metal film is not clear. Glass - fragile amorphous material and the diagram "stress - strain" does not have the field of plastic flow, i.e., it should not have a residual strain [14]. On the other hand, on the surface of glass has a thin damaged layer associated with the manufacture of glass technology, there are various kinds of chemical compounds, oxides. Thus, between the Cu deposited film and the glass an intermediate layer exists, which is more plastic than glass [15]. We can therefore expect that when the load is removed the more ductile layer allows you to save a whole picture of the metal film strain.

In fact, during the deposition of the metal film the thin shell in the form of a hemisphere is obtained. Reflection of incoherent light from the hemisphere gives the converging beam in the center of a sphere of radius R. Evaluation according to the formula for the radius of Newton's rings [11] give values ~ 80 m. However, when scanning the back side of the glass plate we observe the complex interference pattern (Figure 3b), indicating the coherent interaction of light rays from the surface of the hemisphere which is modulated due to glass deformation during of the application of a force. Light and dark interference fringes are in the form of hyperbole and are arranged symmetrically with respect to the point of force application. Externally, they resemble isogyre in photoelasticity method [1, 2]. However, in this case they arise through the interaction of coherent light converging rays reflected from the surface of modulated strain. It is seen that the maximum stress in the center of the plate falls off rather rapidly to its edges. Attraction methods of elastic deformation theory and the comparison with the data obtained from the processing of interference patterns, can serve as a way of determining the stress in dealing with various applications.

5. Conclusion

It is shown that the scanners such as CCD (Charge Couple Device) which use the fluorescent lamps with cold cathode allow the observing the "frozen" interference patterns in thin glass plates in situ after application of a force and a thin metal film deposition. It is proposed to use this technique in the analysis of deformations in the flat solids in the solution of applied problems of elasticity theory.

Acknowledgments

The work was supported by the Russian Science Foundation (project # 16-12-10351).

References

[1] Aleksandrov A Y and Akhmetzyanov M H 1973 *The polarization-optical methods deformed body mechanics* (M.: Nauka)p 576
[2] Pisarenko G S 1981 *Experimental methods for the study of strains and stresses* (Kiev. Naukova Dumka) p 583
[3] Clarke D and Grainger J F 1971 *Polarized Light and Optical Measurements* (Pergamon, Oxford)
[4] Froht M 1950 *Photoelastic* (M: GITL) vol 2 p 486
[5] Trumbachev V F and Katkov G A 1966 Measurement of stresses and strains by photoelastic coating(M .: Nauka)
[6] Teokaris P 1972 *Moire patterns in the study of the deformation* (M.: MIR)
[7] Fink K and Rohrbach H 1961 Measurement of stresses and strainsed N I Prigorovsky (M .: Mashgiz)
[8] Azzam R and Bashar N 1981 Ellipsometry and polarized light(M:. Mir)p 582
[9] Kolokoltsov V N, Borovitskaya I V, Silin P V, Nikulin V Ya and Peregudova E N 2014 *Pulse plasma chemical reactor on the basis of the installation plasma focus Proc. of the VII Int. Symp. on Theoretical and Applied Plasma Chemistry* (Ples,Ivanovo Russia) p 339
[10] Kolokoltsov V N, Degtyarev V F, Silin P V and Nikulin V Y 2015 Application of CCD scanners to analyze the deformation of the glass plate and evaluation of film thickness *Proc. of the VI Int. Conf. "Deformation and destruction of materials and nanomaterials (DFMN-2015)* (Moscow, IMET RAS) p 824
[11] Savelyev I V 1982 *The course of general physics* (M :. Nauka) vol 2 p 496
[12] Timoshenko S P 1972 The course of the theory of elasticity (Kiev, Naukova Dumka) p 508
[13] Landau L D and Lifshitz E M 1982 Electrodynamics of Continuous Media ((M:. Nauka) p 620
[14] Handbook of glass manufacture 1963 ed I I Kitaigorodskii (M.: GIL) vol 1 p 1011
[15] Petrovsky G T and Nersesyants V E 1988 *Nuclear-physical methods of analysis of the surface layers of the glass Glass Physics and Chemistry* **14** 641-65

Generation of sub-gigabar-pressure shocks by a hyper-velocity impact in the collider driven by laser-induced cavity pressure

J Badziak[1], M Kucharik[2], R Liska[2]
[1]Institute of Plasma Physics and Laser Microfusion, 01-497 Warsaw, Poland
[2]Czech Technical University, FNSPE, 160 41 Prague 6, Czech Republik
E-mail: jan.badziak@ifpilm.pl

Abstract. The generation of high-pressure shocks in the newly proposed collider in which the projectile impacting a solid target is driven by the laser-induced cavity pressure acceleration (LICPA) mechanism is investigated using two-dimensional hydrodynamic simulations. The dependence of parameters of the shock generated in the target by the impact of a gold projectile on the impacted target material and the laser driver energy is examined. It is found that both in case of low-density (CH, Al) and high-density (Au, Cu) solid targets the shock pressures in the sub-Gbar range can be produced in the LICPA-driven collider with the laser energy of only a few hundreds of joules, and the laser-to-shock energy conversion efficiency can reach values of 10 – 20 %, by an order of magnitude higher than the conversion efficiencies achieved with other laser-based methods used so far.

1. Introduction

The production of high dynamic pressures using shock waves is a hot topic developed in various branches of research, in particular in high energy-density physics (HEDP) [1], material science [2,3] or inertial confinement fusion (ICF) [4]. The pressures in the multi-Mbar range can be achieved by a large variety of methods employing chemical [5] or nuclear explosions [6], irradiation of a solid target by the laser beam [4,7-10] or the laser-produced X-rays [4,11] as well as the hyper-velocity impact, into a dense target, of a projectile accelerated by a pulsed-power machine [12], a light-gas gun [2,13] or a laser driver [14-19]. However, the pressures in the sub-Gbar or Gbar range are attainable only with laser-based methods or the nuclear explosions (which are forbidden presently). A common drawback of all laser-based methods of the shock generation used so far is the low efficiency of energy conversion from a laser to the shock, $\eta_{ls} = E_s/E_L$, which is below a few percent [4,9,15,18] (E_s is the shock energy and E_L is the energy of a laser beam which generates the shock or produces X-rays or accelerates the projectile). In case of the shock generation by the laser-driven hyper-velocity impact, the main reason for the low η_{ls} value is the low energetic efficiency of projectile acceleration in the commonly used ablative acceleration (AA) scheme, which is $\eta_{acc} = E_p/E_L \sim 0.5 - 5$ % [15-19] (E_p is the projectile kinetic energy). As a result, for the generation of sub-Gbar or Gbar shocks of spatial dimensions enabling their practical usefulness (e.g. studies of equations of state of materials), multi-kJ laser drivers are necessary [15]. Recently, a novel efficient scheme of generation of high-pressure shocks - called the LICPA-driven collider - has been proposed [20,21] (LICPA - laser-induced cavity pressure acceleration [22,23]). In the collider, a projectile (e.g. a heavy disc covered by a low-Z ablator) placed in a cavity is irradiated by a laser beam introduced into the cavity through a hole and then accelerated in a cylindrical or conical guiding channel by the pressure created in the cavity by the laser-produced hot plasma expanding from the irradiated side of the projectile or by the photon (radiation) pressure of the ultra-intense laser pulse trapped in the cavity. The projectile accelerated in the channel impacts into a dense target placed at the channel exit. Due to the impact, the projectile energy and momentum are rapidly transferred to the target and when the projectile velocity significantly surpasses the sound velocity in the target, a high-pressure shock is generated in the target. It has been demonstrated [20,21] that using the LICPA-driven collider, a high-pressure shock can be generated in a solid target with the laser beam energy much lower than that required for producing such a shock with other laser-based methods used so far.

In this paper, the process of shock generation in the LICPA-driven collider is investigated with the use of numerical hydrodynamic simulations. A particular attention is paid to the dependence of shock parameters and the laser-to-shock energy conversion efficiency on the impacted target material and the laser driver energy. It is shown that both in case of low-density (CH, Al) and high-density (Cu, Au) solid targets the shock pressure in the sub-Gbar range can be generated in the target with the laser driver energy of only a few hundreds of joules and the laser-to-shock energy conversion efficiency can reach values above 10 %.

2. Results of numerical simulations

For numerical investigations of shock generation by a hyper-velocity impact in the LICPA-driven collider we used the two-dimensional (2D) hydrodynamic PALE (Prague Arbitrary Lagrangian Eurelian) code [24,25]. The numerical simulations comprised both the projectile acceleration phase and the phase of the projectile impact into the solid massive target placed at the collider channel exit. In particular, they included: the interaction of the laser beam with the projectile ablator and the absorption of the laser radiation in the plasma produced and confined in the cavity, the shock generation in the projectile (by the plasma pressure) and its heating and acceleration in the guiding channel, the interaction of the projectile with the channel wall during the acceleration, the impact of the projectile into the massive target and the generation of the shock in the target and, finally, heating the target by the shock, melting and evaporating of the target material and, as a result, the formation of a crater in the target. The simulations of the acceleration phase were carried out in Lagrangian coordinates applying a mesh moving with the plasma. In turn, in the impact phase, the simulations were performed in the Eulerian coordinates with a static computational mesh as the computational domain in this phase does not change. In this phase, the computations had been conducted for a long time (typically, for ~ 1000 ns) up to the moment when parameters (pressure, temperature, velocity) of the shock, generated shortly after the impact and propagating in the target, became too low to melt the target material, and the dimensions of the crater produced as a result of vaporization and melting of the material stopped to increase. Both in the acceleration phase and the impact phase, the QEOS equations of state [26] were applied to describe thermodynamic properties of materials used in the projectile and the impacted targets. For both phases computations were performed in the cylindrical r, z geometry.

The simulations were performed for the same parameters of the LICPA-driven collider as the ones used in the experiment carried out recently at the PALS laser facility [20]. The cavity and the guiding channel walls were made of gold and the main collider dimensions were as follows: the cavity length $L_c = 0.4$ mm, the channel length $L_{Ch} = 2$ mm, the cavity diameter $D_c = 0.3$ mm, and the channel diameter $D_{Ch} = 0.3$ mm. As a projectile in its initial state we applied the gold disc of the thickness of 2.8 um and the diameter of 0.3 mm (the projectile mass $m_p \approx 4$ µg), covered by the CH ablator of the thickness of 5 µm. The massive targets placed at the collider channel exit were made of Au, Cu, Al or CH. The laser beam parameters on the target (CH ablator) were assumed to be relevant to the parameters of the 3ω PALS laser beam, in particular: the laser wavelength $\lambda = 0.438$ um, the pulse duration $\tau_L = 0.3$ ns, the transverse beam intensity distribution ~ $\exp(r/r_o)^6$ with $r_o = 170$ µm. For the laser energy $E_L = 200$ J, these parameters correspond to the beam peak intensity on the target ~ 10^{15} W/cm^2. The laser beam energy varied within the range 10 – 400 J.

The correctness and accuracy of the PALE code used for the simulations was verified for various laser-based schemes of shock generation [20,27,28], in particular for the LICPA-based collider [20]. Highly positive results of these verifications allow us to believe that numerical results obtained in this paper are correct not only qualitatively but also quantitatively within a factor ~ 1.5 or smaller.

Using the PALE code we examined how parameters and the structure of the shock generated in a massive target by the projectile impact are influenced by the properties of the target material and especially by the material density. Figure 1 presents 2D spatial distributions of the density and the pressure in the impacted target ($z < 0$) made of Au or CH at the moment $t = t_{pmax}$ when the pressure in the target attains the highest value. Although maximum values of these parameters in the generated shock significantly depend on the kind of the material, for all target materials used the spatial shape of the shock is similar – the shock is rather flat and homogeneous within 2r = 200 um The quasi-planar shape of the shock is preserved up to ~ 1 ns after reaching by the shock the highest pressure, and after this period a radial expansion of the shock begins to be significant.

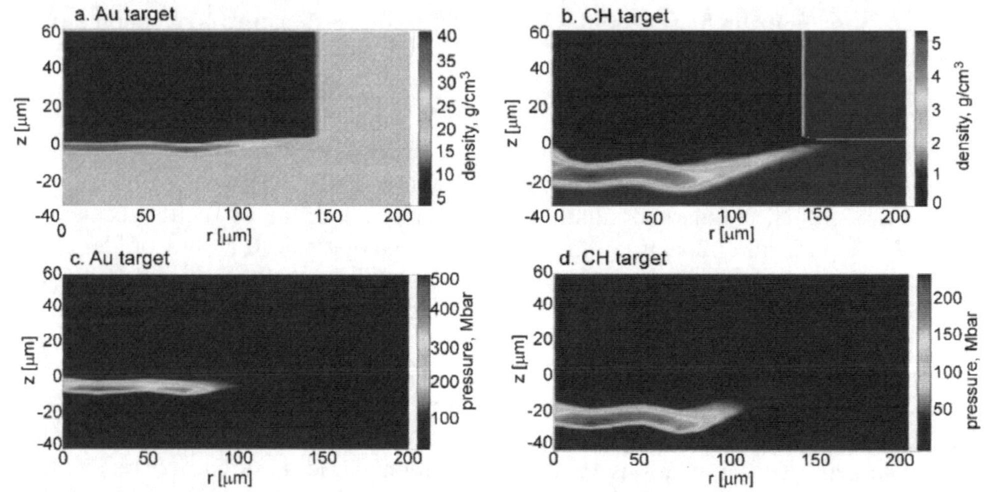

Figure 1. 2D spatial profiles of density (a, b) and pressure (c, d) inside targets ($z \leq 0$) made of Au or CH at the moment when the pressure in the target attains the highest value. The targets are impacted by the gold projectile accelerated in the collider channel situated in the region $z > 0$, $r < 150$ um. E_L = 200 J.

The highest value of pressure in the impacted target changes in time. The simulations showed that the maximum pressure (in the target space) attains its peak value very fast (within ~ 0.2 ns) independently of the target material. It indicates a very short time of energy transfer from the projectile to the target, which is a consequence of the high velocity and compactness (small effective thickness) of the projectile. However the "pressure pulse" duration depends on the target material and for the gold target the duration (FWHM) is by a factor ~ 1.5 shorter than for the plastic target. In the time period of $\Delta t = 1$ ns after the impact, when planarity of the shock is preserved, the maximum shock pressure averaged in time is by a factor 1.5 – 2 lower than the peak pressure and lays in the sub-Gbar range for all the targets used in the simulations. This period is by a factor ~ 2 longer than that estimated for the sub-Gbar pressure range in the hyper-velocity impact experiment with the multi-kJ Nova laser [15] and seems to be long enough to measure some parameters of the material compressed by the shock, e.g. those required for the determination of equation of state of the material.

The summary of quantitative results of the simulation of shock generation by the projectile impact in the targets made of different materials are shown in figures 2, 3 and 4. Figure 2a presents the maximum (in space and time) pressures and velocities of the shocks attained in the targets and figure 2b the laser-to-shock energy conversion efficiency $\eta_{ls} = E_s/E_L$ for the time $t = t_{pmax}$ when the pressure reaches the highest value and for t equal to 5ns after the impact. The shock energy E_s is defined here as the energy of the target material which is flowing along the axis z into the target interior with the velocity higher than ¼ of the maximum flow velocity. The maximum shock velocity decreases with an increase in the target material density while the maximum shock pressure increases with the material density and it approaches 500 Mbar for Au and Cu and surpasses 300 Mbar for Al. So far, such high values of the pressure could be produced with the laser-based methods only with the use of large multi-kJ lasers. Here, the sub-Gbar shock pressures are achieved at laser energy of only 200 J. Achieving such high pressures in the LICPA-driven collider at relatively small laser energy is possible thanks to the very high efficiency of the projectile acceleration by the LICPA mechanism, high efficiency of the energy transfer from the projectile to the target due to high projectile density and a short interaction time between the projectile and the target (being the result of compactness and high velocity of the projectile). The first two factors also determine the high value of the laser-to-shock energy conversion efficiency which for low-density targets (CH, Al) approaches 20 % and is by an order of magnitude higher than the efficiencies attained with laser-based methods so far.

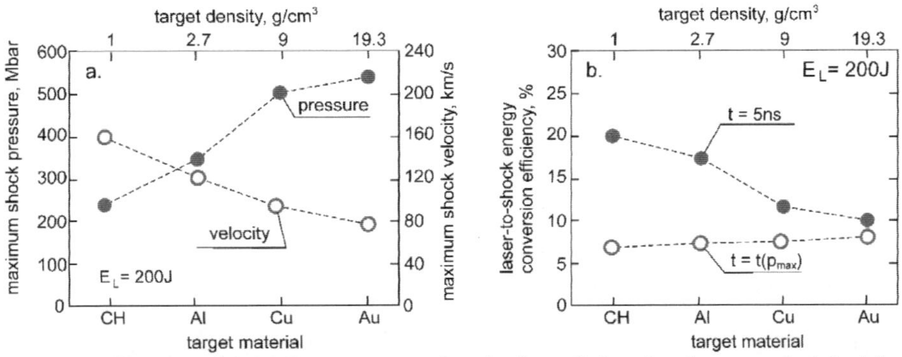

Figure 2. The maximum values of pressure and velocity of the shock generated inside the targets made of various materials impacted by the gold projectile (a). The laser-to-shock energy conversion efficiency for the targets made of various materials at the moment $t = t_{pmax}$ when the shock pressure achieves the highest value and at the moment of 5 ns after the gold projectile impact (b). $E_L = 200$ J.

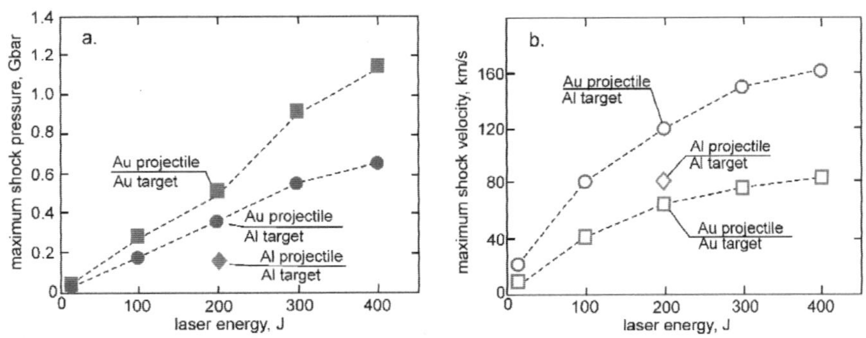

Figure 3. The maximum values of pressure (a) and velocity (b) of the shock generated inside the Au (squares) and Al (circles) targets impacted by the gold projectile as a function of laser energy. The diamonds corresponds to the case when the Al target is impacted by the aluminum projectile of the mass equal to the mass of the gold projectile.

The dependencies of the maximum (in space and time) pressure and velocity of the shocks generated in the Al and Au massive targets by the gold projectile accelerated in the LICPA-based collider, for the energy range of 10 – 400 J accessible at the PALS facility, are presented in figure 3. The shock pressure increases almost linearly with the laser energy for both targets and at $E_L = 400$ J it attains 0.7 Gbar for Al and near 1.2 Gbar for Au. The shock velocity increases approximately as $E_L^{1/2}$ with an increase in the laser energy and for $E_L = 400$ J it reaches 160 km/s for Al and about 80 km/s for Au. In the figure, there is also presented the maximum pressure and velocity of the shock generated in the Al massive target by the Al projectile of the same mass (and areal density) as that of the Au projectile and driven by the laser energy of 200 J. It can be seen that the maximum pressure and the maximum velocity of the shock generated by the gold projectile are greater by a factor of ~ 2 and ~1.5, respectively, than those of the shock produced by the Al projectile. The main reasons for the observed difference in the shock parameters are higher density (by a factor ~ 3 – 4) and smaller thickness of the impacting gold projectile as compared to the Al projectile which allows for the transfer of the gold projectile energy to the impacted target with higher efficiency and during a shorter time than in the case of the Al target.

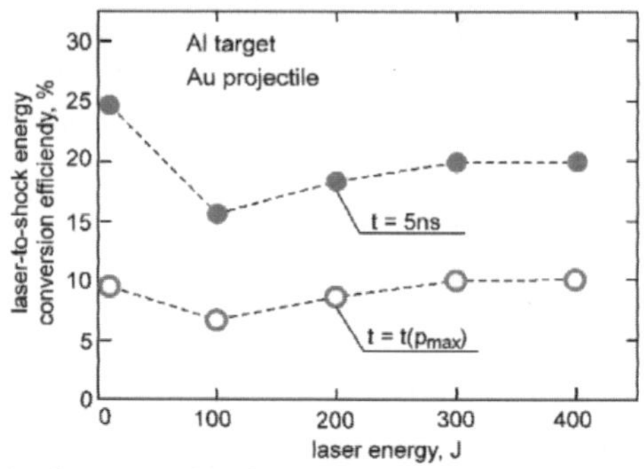

Figure 4. The dependence of the laser-to-shock energy conversion efficiency in the Al target impacted by the gold projectile on the laser energy for the moment when the shock pressure reaches the highest value (open circles) and for the moment of 5 ns after the impact (full circles).

The laser-to-shock energy conversion efficiency, η_{ls} , as a function of the laser energy, for the Al massive target and two moments after the gold projectile impact, is presented in figure 4. The conversion efficiency decreases with an increase in the laser energy in the low energy range (below of 100 J) and then slightly increases. At $E_L = 400$ J, the conversion efficiency reaches 10% at the early phase of the shock formation (when the shock pressure reaches the maximum value) and 20% in the late phase (at t = 5ns). Qualitatively similar dependence of the conversion efficiency on the laser energy was observed for the Au target; however, in this case the conversion efficiency in the high energy range was < 15%.

3. Conclusions

In conclusions, the production of high-pressure shocks in solid targets using the hyper-velocity impact in the recently proposed LICPA-driven collider has been investigated using hydrodynamic simulations. It has been found that with the laser driver energy of only a few hundred joules, sub-Gbar-pressure shocks can be generated in the collider with the laser-to-shock energy conversion efficiency by an order of magnitude higher than in other laser-based schemes used so far. The peak pressure of the shock depends on both the projectile density and the impacted target density and it increases with an increase in these densities. In the laser energy range of 10 – 400 J, the peak pressure approximately linearly increases with the laser energy. The laser-to-shock energy conversion efficiency relatively weakly depends on the laser energy; however, it significantly depends on the target density – for a gold projectile driven by the laser pulse of energy of 200 J it reaches ~ 10 % for the Au target and ~ 20 % for the CH target. For all the target applied (Au, Cu, Al, CH) the pressure in the shock increases to its peak value very rapidly (within $\Delta t < 0.3$ ns for $E_L = 200$ J) and for the time period of $\Delta t = 1$ ns after the impact, when planarity of the shock is preserved, the maximum shock pressure averaged in time lays in the sub-Gbar range both for the high-density and low-density targets used in the simulations. The results presented in the paper suggest that the LICPA-driven collider can be an efficient tool for the shock generation both in case of laser energies higher than those considered in the paper (~ 1 kJ or higher) and in case of low laser energies (~ 5 – 10 J) produced by commercial lasers. The last possibility would open a prospect for conducting research in high energy-density science also in small, university-class laboratories.

Acknowledgments

This work was supported in part by the Ministry of Science and Higher Education, Poland under the grant no W39/7.PR/2015. Also, this work was supported in part by the Czech Ministry of Education, Youth and Sport , projects LD 14089 and RVO 68407700, as well as by the Eurofusion Consortium, project AWP15-ENR-01/CEA-02.

4. References

[1] Drake RP 2006 *High-Energy-Density Physics* (Springer, Berlin).
[2] Nellis W J 2016 *Rep. Prog. Phys.* **69**,1479.
[3] Kraus D, Ravasio A, Gauthier M, et al. 2016 *Nat. Comm.* **7**, 10970.
[4] Atzeni S and Meyer-ter-Vehn J, 2004 *The Physics of Inertial Fusion* (Clarendon, Oxford).
[5] Fortov V E 1982 *Sov. Phys. Usp.* **25**, 781.
[6] Trunin R F 1994 *Phys. – Usp.* **37**, 1123.
[7] Veeser L R and Solem S C 1978 *Phys. Rev. Lett.* **40**, 1391.
[8] Batani D, Morelli A, Tomasini M, et al. 2002 *Phys. Rev. Lett.* **88**, 235502.
[9] Gus'kov S Yu,. Kasperczuk A, Pisarczyk T, et al. 2007 *J. Exp. Theor. Phys.* **105**, 793.
[10] Batani D, Antonelli L, Atzeni S, et al. 2014 *Phys. Plasmas* **21**, 032710.
[11] Löwer Th, Sigel R, Eidmann K, et al. 1994 *Phys. Rev. Lett.* **72**, 3186.
[12] Lemke R W, Knudson M D, Bliss D E, et al. 2005 *J.Appl. Phys*, **98**, 073530.
[13] Mitchell A C, Nellis W J, Mariorty J A, et al. 1991 *J. Appl. Phys.* **69**, 2981.
[14] Obenschain S P, Whitlock R R, McLean E A, et al. 1983 *Phys. Rev. Lett.* **50**, 44.
[15] Cauble R, Phillion D W, Hoover T J, et al. 1993 *Phys. Rev. Lett.***70**, 2102.
[16] Karasik M, Weaver J L, Aglitskiy Y, et al. 2010 *Phys. Plasmas* **17**, 056317.
[17] Frantaduono D E, Smith R F, Boehly T R, et al. 2012 *Rev. Sci. Instrum.* **83**, 073504.
[18] Shui Min, Chu Gen-Bai, Xin Jian-Ting, et al. 2015 *Chin. Phys.* B**24**, 094701.
[19] Min Shui, Genbai Chu, Bin Zhu, et al. 2016 *J. Appl. Phys.* **119**, 035903.
[20] Badziak J, Rosiński M, Krousky E, et al. 2015 *Phys. Plasmas* **22**, 032709.
[21] Badziak J, Krousky E, Kucharik M, and Liska R, 2016 *J. Instrum.* **11**, C03043.
[22] Badziak J, Borodziuk S, Pisarczyk T, et al.2010 *Appl. Phys. Lett.* **96**, 251502.
[23] Badziak J, Jabłoński S, Pisarczyk T, et al. 2012 *Phys. Plasmas*, **19**, 053105.
[24] Liska R, Kucharik M, Limpouch J, et al. 2011, in Fort J, et al., editors, *Finite Volumes for Complex Applications VI, Problemes & Perspectives.* Vol. 2 p. 57-73, Springer-Verlag.
[25] Kapin T, Kucharik M, Limpouch J, et al. 2008 *Int. J. Numer. Methods Fluids* **56**, 1337.
[26] More R M, Warren K H, Young D A and Zimmerman G B 1988 *Phys. Fluids* **31**, 3059.
[27] Koester P, Antonelli L, Atzeni S, et al. 2013 *Plasma Phys. Control. Fusion* **55**, 124045.
[28] Badziak J, Antonelli L, Batani D, et al. 2015 *Laser Part. Beams* **33**, 561.